New Approaches
to Speciation in the
Fossil Record

New Approaches to Speciation in the Fossil Record

edited by

Douglas H. Erwin

and

Robert L. Anstey

Columbia University Press   New York

Columbia University Press
New York   Chichester, West Sussex
Copyright © 1995 Columbia University Press

Library of Congress Cataloging-in-Publication Data

New approaches to speciation in the fossil record / edited by Douglas H. Erwin
  and Robert L. Anstey.
     p.   cm.
  Includes bibliographical references (p. 1–337) and index.
  ISBN 0-231-08248-7 cl
  ISBN 0-231-08249-5 pa
  1. Evolutionary paleobiology.   I. Erwin, Douglas H., 1958–   .
  II. Anstey, Robert L.
  QE721.2.E85N49   1995
  560—dc20   95-7117
                                                                         CIP

Printed in the United States of America

c 10 9 8 7 6 5 4 3 2 1
p 10 9 8 7 6 5 4 3 2 1

Contents

Contributors

Warren D. Allmon, Paleontological Research Institution, 1259 Trumansburg Road, Ithaca, NY 14850

Robert L. Anstey, Department of Geological Sciences, Michigan State University, East Lansing, MI 48824

Gordon C. Baird, Department of Geosciences, SUNY College at Fredonia, Fredonia, NY 14603

Carlton E. Brett, Department of Earth and Environmental Sciences, University of Rochester, Rochester, NY 14627

Alan H. Cheetham, Department of Paleobiology, NHB-121, Smithsonian Institution, Washington, DC 20560

Niles Eldredge, Department of Invertebrates, The American Museum of Natural History, Central Park West at 79th Street, New York, NY 10024

Douglas H. Erwin, Department of Paleobiology, NHB-121, Smithsonian Institution, Washington, DC 20560

Dana H. Geary, Department of Geology and Geophysics, 1215 W. Dayton Street, University of Wisconsin, Madison, WI 53706

Jeremy B. C. Jackson, Center for Tropical Paleoecology and Archaeology, Smithsonian Tropical Research Institute, Box 2072, Balboa, Republic of Panama

Bruce S. Lieberman, Department of Invertebrates, The American Museum of Natural History, Central Park West at 79th Street, New York, NY 10024. Currently at Department of Geology and Geophysics, Kline Geology Laboratory, Department of Geology and Geophysics, P.O. Box 208109, Yale University, New Haven, CT 06520-8109

Charles R. Marshall, Department of Earth and Space Sciences, University of California, Los Angeles, CA 90024

Michael L. McKinney, Department of Geological Sciences and Graduate Program in Ecology, University of Tennessee, Knoxville, TN 37996

Joseph F. Pachut, Department of Geology, Indiana University–Purdue University, Indianapolis, IN 46202

Peter J. Wagner, Department of Geophysical Sciences, University of Chicago, 5734 S. Ellis Avenue, Chicago, IL 60637

Acknowledgments

The editors would like to thank the contributors, particularly those who completed their manuscripts on time, for their patience in a process which took far longer than we originally intended. We would also like to thank those who reviewed the manuscripts in detail, and thereby significantly improved the contributions: Bill DiMichele, Scott Lidgard, Pete Sadler, and some anonymous reviewers.

New Approaches
to Speciation in the
Fossil Record

Introduction

Douglas H. Erwin and Robert L. Anstey

The formation of new species lies at the core of evolution. It is responsible for the incredible diversity of life we find around us and the rich biological complexity documented by the fossil record. Speciation may involve the consolidation of intraspecific adaptive trends, the production of new groups for species sorting, and the establishment of new diversity, whether morphologic, genetic, or behavioral. Yet our understanding of the processes and mechanisms involved in speciation often seem as elusive as they were to Charles Darwin when he wrote *The Origin of Species*.

Evolutionary biologists (among whom we include paleobiologists) have chronicled a multitude of apparent speciation patterns since 1859 and have delved into the ecological, behavioral, genetic, phylogenetic, and philosophical aspects of lineage splitting. From this work generalizations have periodically emerged, often to be tested and found wanting. Despite these difficulties knowledge of speciation has progressed considerably, both in an appreciation of the richness of the process and through the development of new techniques to explore mechanisms.

We were motivated to produce this volume for two reasons. First, paleontology was largely ignored by Daniel Otte and John A. Endler in their 1989 edited volume, *Speciation and Its Consequences*. Yet in our view paleontology can and must play a pivotal role in understanding speciation. Second, we were spurred to action by the fine papers presented at the symposium the two of us organized to commemorate the twentieth anniversary of the 1972 paper by Niles Eldredge and Stephen Jay Gould,

"Punctuated Equilibria: An Alternative to Phyletic Gradualism." This symposium was conducted at the 1992 National Meeting of the Geological Society of America.

Eldredge and Gould's paper sparked a wide-ranging (and often divisive) debate on patterns in the fossil record as well as some fundamental conceptual advances. It also encouraged paleontologists to examine evolutionary patterns with new ideas in mind and to develop new approaches to testing hypotheses with the data of our profession. In organizing the symposium, and this volume, we decided to emphasize newer approaches to analyzing speciation patterns in the fossil record, particularly the contributions of younger paleontologists. We sought to avoid reopening many of the (often fruitless) debates which so exercised evolutionary biologists, sociologists, paleontologists, and others during the first decade following the proposal of punctuated equilibria. So as not to simply revisit the arguments and unresolved debates of the 1970s, we decided to exclude discussions of species selection and many other aspects of the controversy over punctuated equilibria versus phyletic gradualism. As Dana H. Geary observes in her paper in this volume, "After twenty years of empirical investigation into the tempo and mode of species-level change in the fossil record, it is clear that both punctuated equilibria and gradualism occur, as do a variety of intermediate patterns." Even a cursory look at the papers in Otte and Endler's book reveals such a diversity of speciation patterns as to make any hope of a unified paradigm illusory.

Gould and Eldredge provided their perspective on the first twenty years of the idea in a 1993 article published in *Nature*. The current positions of many of the original protagonists are discussed in a 1989 book by Somit and Peterson, which includes contributions on both the biological and sociological aspects of the controversy. As Somit and Peterson observe in their introduction, there remains substantial debate, even among supporters of punctuated equilibria, about the key components of punctuated equilibria: Does the bulk of morphologic change occur during speciation events? What role does selection play in speciation? What are the limits of dynamic stasis between speciation events? Do different modes of speciation dominate in different settings? These are important questions about the tempo and mode of speciation. Answering them (with data rather than polemics) requires well-constructed studies of the fossil record, with consideration of potential sampling problems. Fossil sequences must be selected with care to ensure that the object of study can actually provide useful information about the problem (a step all too often neglected in past studies). Phylogenetic relationships among the species involved must be rigorously documented; the patterns of morphologic change must be studied in detail.

This volume is in three parts. The two-paper "Overview" begins with a historical perspective on the issues considered in this volume, moving on to current controversies and future research directions. Part II, "Speciation Patterns and Processes," includes five papers—all of which touch on the analysis of speciation patterns. These papers are far more than case studies of speciation patterns, for each raises important methodological issues for paleobiologists. The final set of papers, "Macroevolutionary Issues," takes us full circle to the questions raised by Simpson (1944) concerning the connection between patterns in the fossil record and macroevolution.

Stasis and the paradox that morphologic change appears to be closely associated with speciation rather than effectively decoupled from it is addressed by Niles Eldredge in his contribution. Although many authors question the generality of Eldredge's premise (see, for example, Levinton 1988; McKinney and Allmon, this volume; Geary, this volume), the association is sufficiently frequent to make the question an important one (Williams 1992). Eldredge roots his discussion in the works of Simpson, Mayr, and Dobzhansky, arguing that the theory of punctuated equilibria was developed as a bridge between the divergent views of Simpson on the one hand and Mayr and Dobzhansky on the other (see also Eldredge 1985). Stasis is one of the twin pillars of the theory of punctuated equilibria, but in many ways it appears to be the most perplexing. If stasis is real, what happens to all the within-species adaptive changes chronicled by biologists? If stasis can be dynamic or oscillatory, then how can one define its limits? Eldredge reviews a variety of models for long-term stasis, linking speciation and morphologic change, and argues that Mayr's peripheral isolate model of speciation remains the best explanation for the pattern. Finally, Eldredge turns to the issue of species sorting and macroevolutionary pattern. He argues that a new species is usually fairly similar to its parent, but that extinction probability is inversely related to the degree of differentiation. Thus species sorting (not speciation per se) acts as a generator, or ratchet, of adaptive change and of the trends seen in the fossil record. Eldredge suggests that stasis and punctuation may reflect differential probabilities of survival for new species through time.

One question posed by Eldredge is "why do we seem to see a marked bias toward well-differentiated species-level taxa in the fossil record?" Whether this is actually what we see, or simply what we think we see, remains an important question. In a talk at our symposium, Elena Tabachnick (then at the National Museum of Natural History) described her morphometric studies of the Miocene foraminiferan *Globorotalia*, which convincingly demonstrated that "species" boundaries may be far less discrete than often appreciated. Under such circumstances any discussion of patterns of speciation becomes moot: no clearly spatiotemporally bounded individuals exist! The editors regret that Tabachnick was unable to contribute a paper for this volume.

Dana Geary points out that patterns of morphologic change that are not geologically instantaneous may have received too little attention. She describes two lineages of melanopsid gastropods from eastern Europe that exhibit long-term gradual morphologic change over a prolonged period, rather than stasis, and suggests that the data indicate the persistence of relatively constant biological or physical effects over the interval. As she notes, such patterns are as difficult to understand as long-term stasis. During the symposium Peter Sheldon of The Open University discussed similar patterns in Ordovician trilobites. Unfortunately Sheldon was unable to complete his paper for this volume, but a recent statement of his views may be found in Sheldon 1993.

Biological and paleobiological approaches to speciation often seem incommensurate: biologists can examine process but not the temporal dimension; paleontologists often seem left only with pattern, and an imperfect one at that. Several papers in this volume do, however, focus on process rather than pattern. This is an intriguing new trend in paleontology. Previously, practitioners in our field argued that punctuations in the fossil record occur too rapidly to be preserved. Pattern could be described, but because a variety of processes can potentially produce a given pattern, paleontologists were not actually learning more about the process of speciation.

Several contributors here demonstrate that this limitation in paleontology no longer holds. Peter Wagner and Douglas Erwin discuss the relationship between modes of speciation and topologies derived from cladistic analyses of fossils. They investigated the relationship between speciation modes (vicariance, peripheral isolation, etc.), various species-level characteristics (geographic and temporal distributions), and expected clade topologies. While cladistic analyses are clearly insufficient to specify the mode of speciation in particular cases, Wagner and Erwin suggest that species-level clade topologies can be used either to corroborate or refute hypothesized speciation modes.

Michael McKinney and Warren Allmon propose a series of testable models based on metapopulation theory which can be employed directly to analyze speciation in the fossil record. While acknowledging the distortions produced by taphonomic effects, they suggest the record may often not be as bad as commonly believed. They contend that the field of metapopulation dynamics may be harnessed to retrieve greater information about the dynamics of fossil populations (and thus species) than has previously been the case. McKinney and Allmon discuss the relationship between disturbance, at a variety of scales, and diversity, with particular reference to speciation effects. They go on to explore the implications of these ideas for taphonomy, paleoecology, and macroevolution.

Alan Cheetham and Jeremy Jackson also go beyond simply describing pattern to elucidate process in their work on two genera of Tertiary–Recent Caribbean cheilostome bryozoans, *Metrarabdotos* and *Stylopoma*. Their previous work has convincingly demonstrated a pattern of punctuated equilibria for species within these genera and congruence between morphospecies and biologic species (Jackson and Cheetham 1990). In this contribution Cheetham and Jackson show that the topologies obtained by stratophenetics and cladistics are incongruent, and they argue that other evidence provides greater support for the stratophenetic results. More importantly, however, they have employed approaches from quantitative genetics to conclude that stabilizing selection is responsible for long-term morphological stasis within species, and that the brief intervals of speciation may be due solely to mutation and random genetic drift, rather than directional selection.

How much confidence should we place in stratigraphic data? This is the question addressed by Charles Marshall in his contribution. The reliability of the stratophenetic approach of Cheetham and Jackson depends in part on how well the observed pattern of first and last occurrences matches the underlying pattern of origination and extinction. A variety of taphonomic difficulties can play havoc with the record. Paleontologists are thus no longer justified in accepting the record unquestioned. Marshall has developed techniques for estimating the reliability of any particular fossil sequence, and he uses these techniques here to analyze the *Metrarabdotos* data of Cheetham and Jackson. It is important to recognize, however, that Marshall's method assumes randomly distributed fossil horizons—an assumption often violated in the fossil record. Nonetheless, Marshall has made an important contribution by enabling paleontologists to assess the reliability of the record.

Certain generalities emerge from each of these case studies: the need for detailed sampling and careful consideration of the depositional environment, the utility of quantitative treatments of the patterns of morphologic change or stasis, and the importance of rigorous phylogenetic analyses in establishing the phylogenetic patterns. Although more of these elements are present in single studies today than in the past, few studies combine all three, suggesting that one of the greatest hurdles in understanding speciation in the fossil record is the lack of an accepted protocol for such studies. The elements of such a protocol are discussed in numerous papers within this volume, however.

The final three papers do not directly address the controversy over punctuated equilibria or modes of speciation, but rather focus on the implications of speciation patterns for macroevolutionary patterns and paleoecology. Species selection has often been considered the primary generator of macroevolutionary trends and patterns. None of the macroevolutionary analyses in this volume, however, supports species selection within the lineages studied.

The macroevolutionary implications of speciation patterns in Paleozoic stenolaemate bryozoans are discussed by Robert Anstey and Joseph Pachut. They argue that speciation during the early Ordovician was characterized by greater rates of morphologic evolution (quantum evolution) than later in the Paleozoic. The major bryozoan clades, with their defining synapomorphies, all appear during the early Ordovician, but these higher level taxonomic differences were not generated by high rates of speciation, nor by species selection. The same pattern is apparent for the Cambrian Radiation of Metazoa, during which speciation rates were low but the morphologic diversification was high (Valentine and Erwin 1987). Anstey and Pachut's results raise the issue of whether the dominant mode(s) of speciation are different during the initial radiation of a clade or are constant throughout the history of a clade. As several authors in this volume observe, there is considerable room for comparative research on patterns of speciation in different environments, for different types of taxa, and at different times during the history of a clade.

The paleoecological context of speciation has often been ignored. We invited the participation of Carlton Brett and Gordon Baird because of the broader perspective provided by their discussion of rapid community reorganization and speciation in benthic marine communities of the Appalachian Basin during the Early Silurian to mid-Late Devonian. This pattern of coordinated stasis, as Brett and Baird term it, reflects long periods of community stability interrupted by geologically brief episodes of evolution and community reorganization. DiMichele (1994) suggests that such patterns are remarkably widespread in both marine and terrestrial communities and pose a significant challenge to ecologists and evolutionary biologists. Brett and Baird's coordinated stasis is also remarkably similar to Vrba's turnover pulse hypothesis (Vrba 1985), although the explanation they advance differs from Vrba's.

In the final contribution in this volume Bruce Lieberman returns to the prevalence of trends in the fossil record. He employs a detailed phylogenetic analysis of turritellid gastropods to test previous suggestions that species selection is responsible for the differential diversification of nonplanktotrophic versus planktotrophic clades. The concepts of species selection and species sorting are clear outgrowths of the theory of punctuated equilibria—although reliable evidence for species selection (as opposed to sorting) remains sparse. Lieberman's paper, like others in this volume, emphasizes the importance of well-corroborated phylogenies as a basis for analyzing patterns of speciation. The phylogeny demonstrates that nonplanktotrophy developed at least twice within the turritellid gastropod clade. His conclusion thus argues for rejection of the species selection model in this case.

In summary, the collective perspective offered by this book is that species and the speciation process both have a time dimension. The analysis of evolutionary products over geologic time spans clearly separates paleobiological from neobiological understandings of speciation. The contributors in this book all agree that process can indeed be inferred from analysis of fossil products placed in their chronologic and environmental contexts. This book is thus a collective appeal for pluralism in understanding the tempos and modes of actual speciation events: paleontological evidence overwhelmingly supports a view that speciation is sometimes gradual and sometimes punctuated, and that no one mode characterizes this very complicated process in the history of life. Instead, times, places, morphologies, and lineages seem to exert contingent effects on the speciation process; speciations are highly variable, but not nearly as variable as species themselves.

It is clear that a new epistemology of species in time is emerging, with new analytical protocols for recognizing the limits of stasis, the pattern of species in time, the heritability of morphologies, the stratigraphy of patchy habitats, and the phylogenetic effects of variance in speciation. No longer are researchers searching for a dominant mode of speciation. The emerging questions focus instead on the causes of different speciation processes in different habitats, times, morphological complexes, and lineages.

References

DiMichele, W. D. 1994. Ecological patterns in time and space. *Paleobiology* 20:89–92.

Eldredge, N. 1971. The allopatric model and phylogeny in Paleozoic invertebrates. *Evolution* 25:156–167.

———. 1985. *Unfinished Synthesis: Biological Hierarchies and Modern Evolutionary Thought.* Oxford: Oxford University Press.

Eldredge, N. and S. J. Gould. 1972. Punctuated equilibria: An alternative to phyletic gradualism. In T. J. M. Schopf, ed., *Models in Paleobiology*, pp. 82–115. San Francisco: Freeman Cooper.

Gould, S. J. and N. Eldredge. 1993. Punctuated equilibrium comes of age. *Nature* 366:223–227.

Jackson, J. B. C. and A. H. Cheetham. 1990. Evolutionary significance of morphospecies: A test with cheilostome Bryozoa. *Science* 248:579–583.

Levinton, J. 1988. *Genetics, Paleontology and Macroevolution.* Cambridge, U.K.: Cambridge University Press.

Otte, D. and J. A. Endler, eds. 1989. *Speciation and Its Consequences.* Sunderland, Mass.: Sinauer.

Sheldon, P. R. 1993. Making sense of microevolutionary patterns. In D. R. Lees and D. Edwards, eds., *Evolutionary Patterns and Processes*, pp. 19–31. Linnean Society Symposium Volume 14. London: Academic Press.

Simpson, G. G. 1944. *Tempo and Mode in Evolution.* New York: Columbia University Press.

Somit, A. and S. A. Peterson. 1989. *The Dynamics of Evolution: The Punctuated Equilibrium Debate in the Natural and Social Sciences.* Ithaca: Cornell University Press.

Williams, G. C. 1992. *Natural Selection: Domains, Levels and Challenges.* Oxford: Oxford University Press.

Valentine, J. W. and D. H. Erwin. 1987. Interpreting great developmental experiments: The fossil record. In R. A. Raff and E. C. Raff, eds., *Development as an Evolutionary Process*, pp. 71–107. MBL Lectures in Biology, Volume 8. New York: A. R. Liss.

Vrba, E. S. 1985. Environment and evolution: Alternative causes of the temporal distribution of evolutionary events. *South African Journal of Science* 81:229–236.

I

Overview

1

Speciation in the Fossil Record

Douglas H. Erwin and Robert L. Anstey

The processes and patterns of speciation present some of the most intractable problems of evolutionary biology. The process of speciation generally requires too long a period of time to be directly observable by biologists, who can only make inferences from the populational and intrapopulational events they observe, and who must reconstruct past events from the attributes of species inferred to be closely related. Paleobiologists are likewise in an unenviable position: the fossil record is generally too coarse in temporal scale and too limited in geographic coverage to provide a detailed history of speciation events. Fossils also do not preserve all the character states that define species; distinctions between fossil species often reflect subjective judgments by taxonomists. Paleobiological studies of species diversity are routinely based upon substitute knowledge: because of a widespread lack of confidence in the objectivity and robustness of the species record, generic and family diversities often serve as proxies for species diversity. Thus neither neobiologists nor paleobiologists can usually "see" speciation in action; both are left to infer what might have happened from a variety of indirect evidence.

As is often the case in paleobiology, evidence from the fossil record constrains the range of the mechanisms applicable to particular evolutionary case histories but does not specify mechanisms precisely. The simplifying effect of this deficiency is that paleobiology has not needed to cope with such difficult issues as the proper biological definition of species, genetic and developmental constraints, and the relative frequency of particular speciation mechanisms. Such issues, however, ultimately must be addressed in the debate over speciation: the definition of species, for example, plays a crucial role in evaluating the importance of alternative speciation mechanisms. Several recent reviews of speciation have been drawn from the neobiological perspective (Barigozzi 1982; Otte and Endler 1989; Palumbi 1992). The pluralism of these discussions leads to the seemingly inescapable conclusion that there is neither a unified definition of species nor a unified explanation of the mechanisms of speciation; nor does either seem likely in the near future. While such debates fall outside the purview of paleobiology, they at any rate do not discourage paleobiologists from likewise developing a pluralism of viewpoints on speciation.

Paleontologists have, over the last two centuries, developed their own empirical protocols for recognizing fossil morphospecies. These protocols were generally typological and invariant prior to the Modern Synthesis of the 1930s and 40s, becoming populational and incorporating variation afterward. Paleontological protocols have always attempted to mimic contemporary biological concepts. Their chief departure was the recognition (under the aegis of the Modern Synthesis) of the so-called evolutionary species or chronospecies (Simpson 1961), an evolving lineage incorporating temporal variation. The present debate about speciation in the fossil record is a debate about the evolutionary phenomenology of fossil morphospecies. Studies like that of Jackson and Cheetham (1990) may eventually clarify the relationships of fossil morphospecies to living species. Because such knowledge is currently very limited, the debates presented here will focus on fossil morphospecies, their patterns of descent, and their tempos of morphological change.

Paleontologists, taking their cue from Darwin's *Origin of Species*, historically settled into an automatic discussion of the gaps in the fossil record when describing speciation. Species were regarded as the incidental by-products of evolution, and speciation was viewed as a long-term continuous process rather than an episodic one. The incompleteness of the fossil record was deployed as the primary explanation for an inability to document gradual change from one morphospecies into another. Typology partially contributed to this problem, but the eventual study of variation in fossils did little to relieve the emphasis on gaps in the fossil record: even variable morphospecies failed to display gradual, directional change. Today many fossil species are still known from only a very few specimens, from very few localities, and from very few beds. Mere observation of evolutionary tempo depends upon retrieval, and therefore upon abundance in the fossil record. The best known fossil species—i.e., those that are abundant, widely distributed, and stratigraphically long-ranging— seemed to display little net morphological change over their durations. One of the early objections voiced to the *Origin of Species* was that of the English paleontologist John Phillips, who in 1860 observed that most fossil species appeared, persisted through one to three formations, and then disappeared. As Fortey (1988) has noted, Phillips's view has persisted through succeeding generations of stratigraphic paleontologists.

Despite the anecdotal (but taxonomically practical) knowledge that many fossil species persisted unchanged, a textbook widely used for several decades (Moore, Lalicker, and Fischer 1952) still attributed such observations to the incompleteness of the record, and caricatured speciation as a long-term gradual process that could only be glimpsed in the fossil record through the episodic windows provided by preservation and sedimentation. Under the aegis of this dominant textbook, paleontologists continued to name species, but rarely examined speciation per se in the fossil record. Rather, they directed evolutionary research to trends in taxa above the species level. If evolution was a continuous and nonhierarchical process, then observations of it at any level could be cleanly extrapolated to all other levels.

Stratigraphic paleontologists (who enjoyed significant economic support from the petroleum industry) favored index or guide fossils in their investigations: geographically widespread species that displayed little or no change over geologically brief durations. The few studies that had purported to demonstrate gradualism at the species level in the fossil record, thereby "proving" that evolution was paleontologically verified (e.g., Trueman 1922), were falsified by the late 1960s. Workers using less coarse sampling designs, and who were more critical in their analysis of morphology (e.g., Hallam 1962), were responsible for these reappraisals. The concept of chronospecies remained a theoretical construct, but with few substantive examples.

By the late 1960s, with improved time resolution in the stratigraphic record and discovery of new sections filling in many gaps, the view of overwhelming incompleteness in the fossil record was becoming less plausible. Nevertheless, in their otherwise innovative 1971 textbook, Raup and Stanley stated (p. 105), "The large gaps between preserved segments of the lineage provide convenient locations for species boundaries. Thus, nature has greatly reduced the species problem in paleontology."

Punctuated Equilibria: A Revolution?

The theory of punctuated equilibria (Eldredge 1971; Eldredge and Gould 1972; irreverently known as "evolution by jerks" [Turner 1986]) was born as a straightforward attempt to translate into the fossil record Mayr's (1963) hypothesis of peripheral isolates. Eldredge and Gould suggested that peripheral isolates—that is, small populations separated from the main body of the species—would be essentially invisible to the fossil record until (and unless, since many "proto-species" would simply disappear) population size and geographic distribution increased sufficiently to raise the likelihood of preservation. From this perspective, the discontinuities in the fossil record were not artifacts; they were real.

In effect, Eldredge and Gould argued that the prevailing paradigm (gradualism punctuated by stratigraphic gaps) was wrong. If the dominant mode of speciation involves peripheral isolates, as Mayr contended, then the sudden appearance of new species in the record is to be expected *even if the record is highly complete.* Stratigraphic or taphonomic incompleteness would merely accentuate the pattern. In their view, once the population expanded, morphologic stasis would ensue (Gould [1993] notes that Hugo de Vries [1905] also developed a punctuated model of speciation, and coined the term *species selection,* yet advocated a very different process of speciation than did Eldredge and Gould [1972]). Stasis was not promoted out of theoretical concerns, but was based upon an empirical assessment of the actual patterns displayed by known fossil morphospecies.

In their initial proposal of punctuated equilibria, Eldredge and Gould arguably went beyond Mayr's concept of speciation (which did not include stasis). Speciation in their view should no longer be regarded as a long-term process (coextensive with evolution itself) but as an *event* (in the geological sense of "event"). Recognition of speciation as an event effectively decoupled intraspecific evolution from macroevolution, creating two levels of a new evolutionary hierarchy. The claim of morphologic stasis during the duration of a species, while arguably based on the fossil record, implicitly called into question the significance of intraspecific adaptation and variation, except as expressed through the generation of peripheral isolates. It was this claim which was to cause the greatest controversy.

Considerable confusion was generated by the changing nature of the claims for punctuated equilibria between 1972 and the mid 1980s. Ruse (1989) recognizes three phases (parallel to the three stages recognized by Gould [1986]). The first phase is that just described. The second phase linked punctuated equilibria with macromutationist ideas (in the sense of Goldschmidt [1940]). This phase is exemplified by Gould's 1980 paper entitled "Is a new and general theory of evolution emerging?" which proposed evolution as a three-tier hierarchy and which explicitly called into question the efficacy of natural selection. Gould later claimed that this line of argument was developed independently of the theory of punctuated equilibria, a claim which may well be true; nonetheless, many readers missed the distinction. The third phase was marked by a retreat from what Ruse (in our view, fairly) characterizes as the extremism of the second phase and a decline in antiselectionist and macromutationist views. This latest phase also engendered support for stasis from quantitative genetic models, along with concomitant claims that punctuated equilibria had been fully predicted by neo-Darwinism.

The theory of punctuated equilibria involved several assumptions and made several testable claims about the nature of speciation. The first of these assumptions was that reproductive isolation was generally linked to morphologic change. Thus the morphologic transitions observed in the fossil record were taken as a valid proxy for speciation. Second, Eldredge and Gould relied on a particular model of isolation; yet the pattern they described could well be produced by other forms of isolation as well. Indeed, many biological criticisms of the theory focused more on Mayr's model than on empirical patterns from the fossil record. Third, Eldredge and Gould claimed that speciation occurred rapidly. As paleontologists, they meant this in a geological sense, involving tens of thousands of years. Many neobiologists, however, misinterpreted this as a claim for biological rapidity, involving only a few generations. This third claim is often difficult to test because: (1) there are no intermediates available in beds below the base of a punctuated species; (2) the diastem immediately below the base of such a species usually accounts for more time than the beds above or below the diastem; and (3) chronostratigraphic resolution in fossiliferous sedimentary rocks is rarely more precise than ± 500,000 years. Fourth and finally, Eldredge and Gould coupled geologically rapid speciation to a claim of long-term morphologic stasis (which is much easier to document).

As Somit and Peterson (1989) rightly point out, there is considerable disagreement, even among the proponents, about the specific nature of the assumptions and claims that grounded Eldredge and Gould's theory. We can, however, use these four major assertions as a framework in which to consider research on speciation in the fossil record over the past two decades.

The first assumption is among the most troubling. Clearly, if the definition of speciation is the acquisition of reproductive isolation (Mayr 1963), then paleobiologists cannot observe speciation directly. Nor can they recognize sibling species, because the subtle behavioral, genetic, and ecologic clues to their identity are not preserved. Quantum changes in morphology, on the other hand, need not reflect speciation (Palmer 1985). Cladistic definitions of species require that they be recognized from multiple populations; cladists equate speciation with the acquisition of particular derived states in phenotypic characters (which unite the derived populations). If the key character states cannot be preserved, then species cannot be recognized in fossils. With the exception of the ground-breaking work of Jackson and Cheetham (1990 and this volume) and similar studies comparing species in fossil and living material, it is difficult to address the linkage between reproductive isolation and morphological change (but note such a linkage in *Drosophila* in Ayala et al. 1974). Yet as Charlesworth (1990) notes, the theory of punctuated equilibria relies upon some form of reproductive isolation and concomitant ecological opportunities as the major conditions for speciation.

Studies of Darwin's finches in the Galapagos (Grant 1986) and *Drosophila* in Hawaii (Carson and Templeton 1984) demonstrate that the assumption of a linkage between reproductive isolation and morphologic change might in fact be reasonable. Nonetheless, it is this aspect of punctuated equilibria which has received the greatest challenge from population geneticists (Charlesworth, Lande, and Slatkin 1982; Lande 1980; Stebbins and Ayala 1981; Templeton 1981; Turner 1986). Conversely, numerous population geneticists have developed models of genetic change consistent with the tenets of punctuated equilibria (Futuyma 1987,

1989b; Kirkpatrick 1982; Newman, Cohen, and Kipnis 1985). Futuyma (1989a,b) observed that much of this disagreement stems from a division between Fisherian and Wrightian traditions in evolutionary genetics. The former emphasizes gene action and the significance of selection, while the latter places greater importance on epistasis and genetic integration. The legacy of Sewall Wright is thus the tradition underlying punctuated equilibria. The genetic constraints favored by the Wrightian approach provide a compelling argument for the probability of stasis.

A parallel tradition has also emphasized the constraints on morphology associated with developmental integration, and has led (via the debate on speciation) to a renewed paleontological interest in the phenomenon of heterochrony (Gould 1977; McKinney 1988; McKinney and McNamara 1990). The perceived abundance of heterochrony in the fossil record (McNamara 1988; Anstey 1987) provides an underlying support for the probability of stasis maintained by epigenetic and developmental factors. The study of morphogenesis and morphometrics has generally followed traditions such as those of D'Arcy Thompson (1942), Child (1941), Bonner (1974), Olson and Miller (1958), and Waddington (1974). In general, the theory underlying these approaches has deemphasized direct gene action, and has favored more complexly mediated and buffered expressions of genetic change. Students of morphological evolution, however, have not yet succeeded in unifying these approaches into a single theoretical framework. Without such a framework, it is difficult to evaluate the significance of punctuated equilibria, which is, at its core, a statement about *morphological* evolution.

The second assumption, the actual geography of small isolated populations that gave rise to new species, is equally intractable. Taphonomic studies of fossil preservation, now numerous, emphasize that the preservation of even a well-skeletonized organism is indeed a rare event. Preserved fossil populations represent, at best, only submicroscopic percentages of their once-living populations. Sedimentation and preservation of sedimentary beds is a highly episodic process. Populations small enough to produce the geologically rapid morphologic changes called for in punctuations have, therefore, a zero probability of preservation. The taphonomic and stratigraphic window through which paleobiologists might hope to see the biogeography of microdemes—be they sympatric, allopatric, peripatric, or whatever—usually is represented by a single bedding plane or diastem. Both paleo- and neobiologists have constraining data, but both lack convincing empirical demonstrations that any particular model holds for the history of a significant number of species, whether fossil or living.

On the other hand, the fossil morphospecies whose patterns of change have been tracked through multiple time horizons all represent enormously large once-living populations. To be common enough to be sampled in fossil populations in densely spaced strata over a significant time duration means, in most instances, that population densities were comparable to those of the Ordovician bryozoan *Parvohallopora* (the eighth paper in this volume), whose colony fragments likely numbered in the hundreds per square meter in thickets dispersed over 600 km of epeiric seaway; these population densities were maintained over a time interval exceeding seven million years. Enormous quantities of fossil material are available for the Jurassic bivalve *Gryphaea*, for planktonic forams and radiolaria, and for Permian fusulines. Quantitative genetic models in which the likelihood of stasis increases with population size would therefore predict nothing but stasis in such species. Yet the recent empirical observations of paleobiologists (table 1.1) have found what appears to be a much richer variety of phenomena.

The third assumption inherent in the punctuated equilibria theory involves the rapidity of speciation. The distribution of morphologic change through time and space remains a compelling question, regardless of one's views on the linkage between morphologic change and the onset of reproductive isolation. Moreover, paleobiologists have the only long-duration data applicable to the question. These data constitute a rich body of empirical knowledge on the temporal dynamics of morphological evolution. Since 1972 numerous paleontological studies have explicitly investigated the tempo of morphological change in the fossil record, with varying results. Many of the observed patterns, such as the maintenance of long-term sustained gradualism (the third paper in this volume), or the even more problematic oscillations between stasis and gradualism, may not yet have an adequate theoretical basis. Calculated selection coefficients seem to be too weak to justify conventional natural selection as a cause for these long-term patterns (Lande 1976). Problems of temporal scaling are also involved, as discussed by Gingerich (1983). The results so far ostensibly support a pluralism of tempos and modes in the historical

Table 1.1

Varieties of Tempos and Modes of Speciation

Mode	TEMPO		
	Gradualism	Gradualism and Stasis	Punctuation and Stasis
	Ordovician bivalves: Bretsky & Bretsky 1977	Ordovician bryozoans: Brown & Daly 1985	Devonian corals: Pandolfi & Burke 1989
	Permian fusulines: Ozawa 1975	Ordovician trilobites: Cisne et al. 1982 Sheldon 1987; 1993	Devonian brachiopods: Isaacson & Perry 1979
	Permian bryozoans: Pachut & Cuffey 1991	Jurassic ammonites: Raup & Crick 1981; 1982	Jurassic bivalves: Johnson 1985
	Jurassic bivalves: Fortey 1988	Cretaceous bivalves: Geary 1987	Cretaceous echinoids: Smith & Paul 1985
	Paleogene primates: Godinot 1985	Cenozoic ostracods: Benson 1983	Eocene condylarths: West 1979
	Neogene radiolarians: Baker 1983	Neogene radiolarians: Kellogg 1985; 1983 Sachs & Hasson 1979	Neogene echinoids: McNamara 1990
Nonbranching	Neogene forams: Malmgren & Kennett 1981 Arnold 1983 Banner & Lowry 1985	Neogene forams: Scott 1982 Malmgren et al. 1983 Malmgren & Berggren 1987 Wei & Kennett 1988	
	Neogene bivalves: Hayami & Ozawa 1975	Neogene fish: Bell et al. 1985	
	Neogene gastropods: Geary 1990; this volume		
	Neogene echinoids: McNamara 1990		
	Neogene rodents: Chaline & Laurin 1986		

(continued)

Table 1.1
Continued

Mode	TEMPO		
	Gradualism	Gradualism and Stasis	Punctuation and Stasis

	Jurassic ammonites:
	  Callmon 1985
	Eocene primates:
	  Gingerich 1976; 1985
		Ordovician trilobites:
		  Fortey 1985
		Cretaceous ammonites:
		  Reyment 1975
			Devonian brachiopods:
			  Goldman & Mitchell 1990
			Cretaceous ostracods:
			  Reyment 1982
			Cenozoic ostracods:
			  Whatley 1985
			Eocene ostracods:
			  Reyment 1985
			Neogene diatoms:
			  Sorhannus 1990a,b
			Neogene radiolarians:
			  Kellogg & Hays 1975

Limited Branching			

(continued)

Table 1.1
Continued

Mode	TEMPO		
	Gradualism	Gradualism and Stasis	Punctuation and Stasis

	Eocene condylarths:
	  Gingerich 1976; 1985
	Eocene primates:
	  Bown & Rose 1987
			Cambrian trilobites:
			  Robison 1975
			Ordovician trilobites:
			  Fortey 1974
			  Henry & Clarkson 1975
			Devonian stromatoporoids:
			  Fagerstrom 1978
			Devonian brachiopods:
			  Fagerstrom 1978
			Devonian trilobites:
			  Eldredge & Gould 1972
			Jurassic bivalves:
			  Hallam 1982
			Cenozoic bivalves:
			  Stanley & Yang 1987
			Neogene radiolaria:
			  Lazarus 1983; 1986
			Neogene bryozoans:
			  Cheetham 1987; this
			  volume
			Neogene molluscs:
			  Eldredge & Gould 1972
			  Williamson 1981
			  Kelley 1983
			Neogene ungulates:
			  Vrba 1980

Multiple Branching			

NOTE: From non-anecdotal paleontological studies published between 1972 and the present volume. Position in the table generally reflects the conclusions of the authors cited, and not subsequent debates.

record of speciation. No single pattern appears dominant, and their variety suggests that new research agendas are needed to investigate the conditions under which any one of them might be predicted. The results shown in table 1.1, representing studies of 58 lineages published during the last 22 years, collectively provide a very different result from the initial predictions of Eldredge and Gould (1972). The greatest departure may be in the preponderance of studies illustrating *both* stasis and gradualism in the history of a single lineage.

The fourth issue raised by the theory of punctuated equilibria concerns morphologic stasis. A principal claim of Eldredge and Gould (1972) is that "stasis is data." Many illustrations of both gradual change (gradualism) and stasis are overly simplistic. In several instances, the same data have even been interpreted as gradualism by one set of authors and stasis by another; Gould and Eldredge (1977) review these instances. In neither tempo should one expect (in continuously varying characters) absolute stasis, nor a directional trend without minor reversals (Sheldon 1993). Distinguishing between ecophenotypic effects, selection, and random walks is difficult in most cases (contra Bookstein 1988). Sequences of short duration that can not be distinguished statistically from the null hypothesis of a random walk represent neither stasis nor gradualism; the tempo in those instances is simply unresolved. Complex forms of stasis and gradualism may be difficult to distinguish. For example, is dead-end gradualism (i.e., gradual change in primitive states or in states *not* leading to a species difference) really another form of stasis? Is gradual change with a very low slope coefficient another form of stasis (Gingerich 1984)? The oxymoron of oscillatory stasis (the second paper in this volume) requires some robust formal protocol that can distinguish it unambiguously from both random walks and reversing gradualism. Short-term clines, for example, might represent a short segment of oscillatory stasis rather than gradualism (the eighth paper in this volume). Multiple forms of both gradualism and stasis have obscured the once-simple distinction between the two, and the issue of temporal scale (Gingerich 1983) confounds the issue even more.

Characters with nominal and ordinal states—for example, those with discrete states, four toes v. five toes—have a tough time displaying gradualism (except as shifting frequencies, a condition rarely ascertainable from the fossil record). These are the characters most commonly used in cladistic analysis, and this style of cladistics assumes punctuation at the character level. Such characters may, however, represent phenotypic thresholds that come to expression following the buildup of genotypic change. The underlying genetic changes could thus well be gradualistic; only the morphologic expression is punctuated.

Most of the studies in table 1.1 have been based upon continuously varying characters—that is, shapes, angles, etc. These are intrinsically the characters most likely to display temporal variation. In such characters some gradualism is to be expected. There is simply no such thing as complete stasis in features, such as shapes, measured on a continuous scale. Such characters are also strongly subject to both ontogenetic and eco-phenotypic variation.

Underlying these four major concerns is a still more pernicious issue: perceptual bias. Sheldon (1993) and Fortey (1988) have emphasized the influence of subtle perceptual biases on the morphologic patterns an observer describes. The first major section in Eldredge and Gould 1972 ("The cloven hoofprint of theory") deals with this issue. Perceptual bias is easiest to discern in studies of tempo and mode in fossils that have been based on single characters (Cheetham 1987) and do not reflect a whole-organism (i.e., phenetic) tempo and mode. For example, of the nine characters measured by Ozawa (1975) for Permian protozoan fusulines, only four have been judged by Gould and Eldredge (1977)—correctly, in our view—to display gradualism. Ozawa, nevertheless, held the opposite, and textbooks still routinely characterize the Ozawa study as outstanding support for gradualism (see table 1.1).

Stratigraphic position in a local section may not accurately reflect phylogeny (the seventh paper in this volume), just as cladograms cannot be automatically converted to phylogenetic trees (the fourth and eighth papers in this volume). One of the problems involved in using cladistics to investigate tempo and mode in speciation is that whole-organism or "phenetic tempo" may become methodologically unobservable. Naturally, the primitive states lingering within any lineage are much more likely to display stasis than are the derived states, and phenetic approaches will automatically find more stasis. Conversely, recognition of "cladistic tempo" may be based on only a small number of derived states. Finally, the multivariate conventions for recognizing phenetic morphospecies (the sixth paper in this volume) may incorporate perceptual biases that may be difficult to detect casually, thus intrinsically biasing the results. Canonical multivariate techniques, for example, are designed to minimize within-group variation and maximize across-group differences. It would be unfair to delineate species by such a technique and then "discover" that their canonical variates display stasis.

Multivariate techniques such as factor analysis are also morphological filters that can reduce many variables to a small number of independent axes. Such axes, however, will reflect the covariance in two or more traits. Plotting factor loadings stratigraphically may be desirable, but it will automatically reduce some covarying traits to stasis. How would the conclusions regarding tempo in Eocene mammal evolution, for example, change if the covarying lengths and widths of their first molars were plotted stratigraphically as a reduced first axis?

Likewise, results can be biased by data transformations. The Eocene dental data used by Gingerich (1976, 1985) and by West (1979) were logarithmic transforms of the product of molar length times molar width. The product alone represents a geometric amplification of the raw measurements. The log transform specifically expands the smaller end of the scale, so that the down-axis trends in *Hyopsodus* would virtually disappear on an untransformed and unmultiplied scale. The raw data must have had disturbingly little variation, and it makes us wonder if the alleged primate *Hyopsodus* might instead have had a cloven hoof. Both cladists and pheneticists should become aware that their standard manipulations of data might render biased conclusions regarding the tempo of morphological evolution.

Species Selection and Sorting as an Explanation for Trends

The theory of punctuated equilibria has a number of implications for larger scale, or macroevolutionary, patterns in the fossil record. Eldredge and Gould (1972) reflected on the origin of trends, but as they later acknowledged (Gould and Eldredge 1977, 1993; Gould 1982, 1989) they did not at first fully appreciate the macroevolutionary implications of their theory. Stanley (1975) articulated the concept of species selection, while Vrba (1980; see also Vrba and Gould 1986) drew the important distinction between species selection and species sorting. In this distinction, species selection involves only the emergent properties of species, not the adaptations of individuals. Sorting, on the other hand, does not imply selection for group attributes. In time, three postulates were developed for sorting as an explanation of trends: If temporal intraspecific variation is low, as demanded by punctuated equilibria, trends generally arise from a changing mix of species, rather than via anagenetic changes summed across species. Second, sorting operates through the differential origin or extinction of species. Third, if the punctuated view of speciation is adopted as dominant, it favors a hierarchical view of evolution and a rejection of extrapolationism across levels (Gould 1980, 1993).

A recent consensus on evolutionary trends (McNamara 1990) suggests that there is no general relationship between phyletic trends and trends at the species level and above. The sole exception noted by McNamara was the existence of paedomorphoclines—trends in which later adults take on progressively more juvenile features—in Paleozoic bryozoans, both within and across species (Anstey 1990; Pachut 1989). In the example cited, paedomorphoclines accounted for 14 of 17 documented heterochronic trends (Anstey 1987). Bryozoan paedomorphoclines might represent, therefore, one of the major exceptions to species sorting as a source of trends.

Many perceptual problems also exist in recognizing macroevolutionary trends in the fossil record, and the same difficulties in epistemology confuse the recognition of trends as confuse the recognition of tempos and modes of speciation. As Gould (1990) puts it, "things are seldom what they seem." Multiple definitions and research protocols are evident in recent research on evolutionary trends (McNamara 1990). For example, most of trends reported in bryozoan evolution fail to meet the standard concept of trend (Anstey 1990); they represent instead clade replacements (without morphological changes within either clade), diversity changes (such as the diversification of a clade or pseudoclade), or temporal changes in polyphyletic (i.e., artificial) taxa. Many reported trends are nothing more than changes in the diversity of a lineage (automatically shifting the maximum state acquired as diversity changes). True morphological trends, of which very few are known, involve a sustained, statistically significant, temporal shift in the *total* morphological range of a monophyletic lineage. Further analysis of trends is needed, coupled with analysis of the tempo and mode of the speciations within the trend. Only then can such mechanisms as species sorting be reasonably evaluated.

Current Status

Little doubt exists in anyone's mind that species must be studied in both time and space, over their entire history and geographic spread. Ideally, species must be placed accurately within their phylogenetic lineages, and their actual ancestors (rather than sister groups) must be known. (For how many lineages can this claim be made?). Detailed sampling within the boundary interval between an ancestral and a descendant species might reveal the tempo of speciation as well as its geographical pattern. Studies that have so far attempted the historical analysis of tempo and mode have yielded an apparent plurality of results: we compiled table 1.1 from only nonanecdotal studies that provided continuous morphologic records reflecting change both within and among species over a definable stratigraphic interval. We make no further claims for the validity of the tempos and modes reported (several of which have been debated), noting substantial variation in the number of species surveyed, the number of characters measured, the basis of phylogenetic assertions, the durations of the stratigraphic windows, the geographic coverage, and the analysis of stratigraphic gaps and time resolution.

An interesting departure in table 1.1 from conventional expectations is that fully one-fourth of the studies we assessed describe single lineages that sequentially or simultaneously display both gradualism and stasis—more than half of which represent metazoans. Several other studies could be relocated to this column as well. For example, the Eocene mammalian condylarths that display multiple branching and gradations in one region (Gingerich 1976, 1985) display no branching and stasis in an adjacent region (West 1979). The Permian fusulines studied by Ozawa (1975) evolved gradually in four characters, but not in five others; therefore they might also be moved to the center column. Similarly, the short-term microevolution in Neogene bivalves (Hayami and Ozawa 1975) and in Permian bryozoans (Pachut and Cuffey 1991) might easily be interpreted as just short segments of oscillatory stasis or random walks (therefore they might land in any column). If some of the punctuated patterns fail to hold up—such as those observed in Cenozoic molluscs by Williamson (1981), which might be attributable to ecophenotypic responses to increased salinity during a lacustrine regression (Kat and Davis 1983)—then the gradation-and-stasis column might come to hold as many empirical examples as does the punctuated pattern. This possibility is not so surprising, because all of the taxa displaying gradualism in the particular characters studied by the researchers in this table must simultaneously display stasis in others, namely those characters that define and delineate their lineage. Even the most famous example of natural selection and gradualism—the peppered moth, *Biston betularia*—displays gradualism only in its protective coloration, and ostensibly stasis in all other features. Caution is also required in proclaiming sequential alternations of gradualism and stasis, because varying sedimentation rates could generate condensed and diluted intervals in succession that falsely appear as rapid gradualism and stasis (MacLeod 1991).

The study by Brown and Daly (1985), while not reaching any explicit conclusions regarding tempo and mode of evolution, is a highly informative example. In three species of the Ordovician bryozoan *Parvohallopora,* the characters defining each species lineage were invariant over a duration exceeding many millions of years (see the eighth paper in this volume), and thus represent stasis. Many other characters were measured, however. Using a statistical technique that reduces nonheritable variation (see the eighth paper), all were plotted stratigraphically and tested for linear trends. Both stasis and sustained trends appear in some characters in some stratigraphic sections, but not in other characters or in the same character in other sections. Some sections display short-term reversing clines in some characters. Therefore these lineages display stasis, oscillations or random walks, and local but not regional gradualism in a mosaic of biometric characters. If the study had included fewer lineages, fewer characters, fewer sections, sampling over a more limited interval, or if it had failed to reduce ecophenotypic variation, then the conclusions might well have been different, as they might also have been if only selected results had been published.

Brown and Daly (1985) and Jackson and Cheetham (1990) made efforts to account for the effects of heritable versus nonheritable variation (see also Pachut 1989). Hallam (1982) similarly removed the effects of ontogenetic variation from the evolution of the Jurassic bivalve *Gryphaea*. No explicit applications of the technique of fluctuating asymmetry ("FA" of Palmer 1986) have yet been made in the analysis of tempo and mode, despite FA's apparent correlation with genetic heterozygosity; this technique clearly deserves more attention. In general, more attention should be given to the variation genetics of fossil populations (Van Valen 1969). Best (1961), for example, claimed that the distribution of three phenotypes in two populations of a Silurian trilobite indicated that these populations were in Mendelian equilibrium. Surely there is still much to come from the study of genetics in the fossil record and its potential bearing on the speciation debate.

More than one-fourth of the studies in table 1.1 are based on protozoans, which may not form species comparable to those in outbreeding metazoans (Tabachnik 1992). Protozoan lineages, rather, may alternate between sexual and asexual reproduction for unknown durations. The onset of environmental stress might trigger more widespread sexual reproduction, possibly producing the alternations between stasis and gradualism seen in some lineages. Many of these alternations, however, surely reflect unrecognized hiatuses in deep-sea sediments (MacLeod 1991). The increase in the size of the initial chamber in Permian fusulines (Ozawa 1975) might likewise reflect an extremely sluggish decrease in outbreeding during the Permian (the initial chamber is normally larger in asexually produced generations). Overall, however, the tempos and modes of protozoan evolution show patterns that have many parallels among metazoans.

The asymmetry of table 1.1 suggests that nonbranching lineages are more likely to display either gradualism or gradualism-and-stasis, and that highly branched lineages are more likely to display stasis. Therefore the factors that can increase the frequency of branching might also lead to more stasis. The factors that produce less branching could lead to more gradualism. This dichotomy is similar to that predicted by Sheldon (1993), but differs from his formulation because stable environments often maintain higher diversities than unstable ones, and offer more opportunities for lineage branching. The frequency of branching also differs strongly among clades, from diverse and volatile clades such as ammonites, to clades of "living fossils" such as lingulate brachiopods. Branching and vicariance can also result from the intrinsic heterogeneity and degree of isolation among habitat patches. Whatever factors might be controlling branching frequency might also control evolutionary tempo. Another critical issue is the degree of punctuation involved in each branching event: strongly punctuated branching could reduce selection pressures and eliminate subsequent gradualism. An unresolved issue is the degree or quantum level of punctuation achieved in each branching event.

Stratigraphic Resolution and Acuity

Prior to 1980 paleontologists generally assumed that the fossil record was sufficiently complete to study both ecological and evolutionary processes. In other words, there were no inherent or systematic biases known in the stratigraphic record that should limit the scope of the questions paleontologists could address. Thus during the 1970s paleoecologists attempted to describe ecological processes within fossil communities without asking whether or not the record was sufficient to resolve time to the scale of tens or hundreds of years. Other paleontologists, too, reasonably assumed that there was no bar to analyzing the process of speciation and that observed patterns of divergence accurately recorded speciation events.

These assumptions came to a screeching halt in 1980. Papers by Schindel (1980, 1982) and Sadler (1981) demonstrated that in most circumstances marine paleontologists simply could not resolve time finely enough to observe ecological events over a long duration, and raised substantive questions about the ability to resolve fine-scale evolutionary processes. In brief, the difficulty is that deposits which record brief intervals of time (event beds, for example) are too widely spaced in the record to provide continuous information on ecological and speciation events. Yet other deposits are so condensed and/or bioturbated that fine-scale temporal information has been lost. Interestingly, such is not the case with terrestrial megafloral deposits, which are often limited to the leaves deposited in a single year immediately around the locality (Burnham, Wing, and Parker 1992).

The increasing use of phylogenetic (cladistic) methods in paleontology promises more rigorously constrained hypotheses of ancestor-descendant and sister-group relationships. In many groups however, homoplasy is so pervasive that stratigraphic information plays an increasingly important role in testing the validity of phylogenetic hypotheses. Marshall's (1990) development of techniques to approximate confidence intervals on the stratigraphic ranges of species shows great promise. At present Marshall's techniques (see the seventh paper in this volume) assume that fossils are randomly distributed within an outcrop, which is not likely to be generally true, but nonetheless this line of investigation should prove enormously profitable as some species can be excluded as potential ancestors (or descendants) on stratigraphic grounds.

Additional complexity is added by taphonomic processes. For example, time-averaging produces assemblages which combine fossils over a long span of time, removing seasonal and yearly fluctuations and potentially reflecting more stable, long-term community structure. Large storms may rework thousands of years of sediment into a single time-averaged event bed. Longer-term time averaging, produced for example by condensed sections, combines specimens that lived at very different times. Burrowing may partially or completely rework the stratigraphic order of fossils, further destroying the utility of the fossil record for some types of studies.

For speciation studies, time averaging and other taphonomic processes generally limit paleontologists' abilities to reliably discern microevolutionary patterns. As much as we all may want to believe in the evidence from the fossil record, recent studies in taphonomy suggest that we must never overlook the null hypothesis that any particular sequence is insufficient to record ecological and short-term evolutionary processes. Certainly there are stratigraphic intervals where sedimentation rate is high enough, preservation is good enough, and sampling is sufficient that intraspecific events can be chronicled. This, however, is not the null hypothesis, and such circumstances must be carefully documented by the investigator. As to the hope that we might actually be able to follow the process of speciation in the fossil record, that largely depends upon the rate at which speciation occurs in a particular setting.

The Way Forward

One flaw in the current paleobiological approach to finding examples of and better understanding the tempo and mode of evolution may be the lack of an explicit protocol for the investigation of speciation (as well as species trends) in the fossil record. The reliability of studies might increase considerably if investigators were to specify, in advance, the temporal resolution required of stratigraphic sections in which they proposed to study speciation, as well as the data required to demonstrate any of the competing models. All temporal patterns should be tested against the null hypothesis of a random walk (random walks should be discarded as data: they do not reliably represent either stasis or gradualism). The protocol must minimally include abundant individuals from closely spaced stratigraphic intervals, fine-scale temporal resolution, a high degree of phylogenetic resolution, sampling from sections distributed over the geographic range of the taxa studied, biometric capture of morphology with removal of ecophenotypic variation (i.e., assessing heritability; the eighth paper in this volume), and ideally assessments of population genetics (e.g., Palmer 1986). Very few or none of the studies listed in table 1.1 meet the demands of such a protocol.

Work must also continue on the distribution of speciation tempos and modes for different environments, times, life history strategies, and geographic settings. Vrba's (1985) turnover-pulse hypothesis, for example, related punctuated speciation to rapid climatic fluctuations. Sheldon (1993; see also Johnson 1982) has suggested that narrowly fluctuating, stable environments favor phyletic gradualism, whereas unstable environments yield stasis, as patterns supported by a number of empirical studies. Anstey (1987) predicted oscillatory stasis associated with paedomorphosis in low diversity habitats, but also recognized numerous paedomorphoclines in which there was a high probability that intraspecific heterochrony was driving such trends (Anstey 1990; Pachut 1989). Speciation processes might also vary within clade histories (the eighth paper in this volume), and within Phanerozoic history in general, such as between major radiations and recoveries (Stanley 1979) and the history of post-extinction recoveries. Brett and Baird (the ninth paper in this volume) suggest that coordinated stasis generally characterizes the longer intervals between biotic reorganization events. If punctuation characterizes events of greater biological significance, and gradualism and/or stasis characterizes the times between such events, then the concept of a multitier evolutionary hierarchy will be reinforced.

Some final questions could be asked: How much of evolution is explained by speciation, and how much is not? Speciation seems to be anything but a uniform process producing uniform products. Is speciation itself a hierarchy of phenomena, is it the boundary between two tiers of the evolutionary hierarchy, or are species the incidental by-products of a generalized and nonhierarchical evolutionary process? We need more answers.

References

Anstey, R. L. 1987. Astogeny and phylogeny: Evolutionary heterochrony in Paleozoic bryozoans. *Paleobiology* 13:20–43.
———. 1990. Bryozoans. In K. J. McNamara, ed., *Evolutionary Trends,* pp. 232–252. London: Belhaven Press.

Arnold, A. J. 1983. Phyletic evolution in the *Globorotalia crassaformis* (Galloway and Wissler) lineage: A preliminary report. *Paleobiology* 9:390–397.

Ayala, F. J., M. L. Tracey, D. Hedgecock, and R. C. Richmond. 1974. Genetic differentiation during the speciation process in *Drosophila. Evolution* 28:576–592.

Baker, C. 1983. Evolution and hybridization in the radiolarian genera *Theocorythium* and *Lamprocyclas. Paleobiology* 9:341–354.

Banner, F. T. and F. M. D. Lowry. 1985. The stratigraphical record of planktonic foraminifera and its evolutionary implications. In J. C. W. Cope and P. W. Skelton, eds., *Evolutionary Case Histories from the Fossil Record. Special Papers in Palaeontology* 33:117–131.

Barigozzi, C., ed. 1982. *Mechanisms of Speciation: Progress in Clinical and Biological Research.* No. 96. New York: Liss.

Bell, M. A., J. V. Baumgartner, and E. C. Olson. 1985. Patterns of temporal change in single morphological characters of a Miocene stickleback fish. *Paleobiology* 11:258–271.

Benson, R. H. 1983. Biomechanical stability and sudden change in the evolution of the deep-sea ostracode *Poseidonamicus. Paleobiology* 9:398–413.

Best, R. V. 1961. Intraspecific variation in *Encrinurus ornatus. Journal of Paleontology* 35:1029–1040.

Bonner, J. T. 1974. *On Development.* Cambridge: Harvard University Press.

Bookstein, F. L. 1988. Random walks and the biometrics of morphologic characters. *Evolutionary Biology* 23:320–325.

Bown, T. M. and K. D. Rose. 1987. Patterns of dental evolution in Early Eocene anaptomorphine primates (Omomyidae) from the Bighorn Basin, Wyoming. *Memoir of the Paleontological Society* 23:1–162.

Bretsky, S. S. and P. W. Bretsky. 1977. Morphologic variability and change in the paleotaxodont bivalve mollusk *Nuculites planulatus* (Upper Ordovician of Quebec). *Journal of Paleontology* 51:256–271.

Brown, D. G. and E. J. Daly. 1985. Analysis of evolution of structural characters of *Parvohallopora* Singh from the Dillsboro Formation, Cincinnatian Series, Indiana. In C. Nielsen and G. Larwood, eds., *Bryozoa: Ordovician to Recent,* pp. 51–58. Fredensborg, Denmark: Olsen and Olsen.

Burnham, R. J., S. L. Wing, and G. G. Parker. 1992. The reflection of deciduous forest communities in leaf litter: Implications for autochthonous litter assemblages from the fossil record. *Paleobiology* 18:30–49.

Callmon, J. H. 1985. The evolution of the Jurassic ammonite family Cardioceratidae. In J. C. W. Cope and P. W. Skelton, eds., *Evolutionary Case Histories from the Fossil Record. Special Papers in Palaeontology* 33:49–90.

Carson, H. A. and A. R. Templeton. 1984. Genetic revolutions in relation to speciation phenomena: The founding of new populations. *Annual Review of Ecology and Systematics* 15:133–164.

Chaline, J. and B. Laurin. 1986. Phyletic gradualism in a European Plio-Pleistocene *Mimomys* lineage (Arvicolidae, Rodentia). *Paleobiology* 12:203–216.

Charlesworth, B. 1990. Speciation. In D. E. G. Briggs and P. R. Crowther, eds., *Palaeobiology: A Synthesis,* pp. 100–106. London: Blackwell Scientific.

Charlesworth, B., R. Lande, and M. Slatkin. 1982. A neo-Darwinian commentary on macroevolution. *Evolution* 36:474–498.

Cheetham, A. H. 1987. Tempo and mode of evolution in a Neogene bryozoan: Are trends in single morphologic characters misleading? *Paleobiology* 13:286–296.

Child, C. M. 1941. *Patterns and Problems of Development.* Chicago: University of Chicago Press.

Cisne, J. L., G. O. Chandlee, B. D. Rabe, and J. A. Cohen. 1982. Clinal variation, episodic evolution, and possible parapatric speciation: The trilobite *Flexicalymene senaria* along an Ordovician depth gradient. *Lethaia* 13:325–341.

De Vries, H. 1905. *Species and Varieties: Their Origin by Mutation.* Chicago: Open Court.

Eldredge, N. 1971. The allopatric model and phylogeny in Paleozoic invertebrates. *Evolution* 25:156–167.

Eldredge, N. and S. J. Gould. 1972. Punctuated equilibria: An alternative to phyletic gradualism. In T. J. M. Schopf, ed., *Models in Paleobiology,* pp. 82–115. San Francisco: Freeman, Cooper.

Fagerstrom, J. A. 1978. Modes of evolution and their chronostratigraphic significance: Evidence from Devonian invertebrates in the Michigan Basin. *Paleobiology* 4:381–393.

Fortey, R. A. 1974. The Ordovician trilobites of Spitsbergen. I. Olenidae. *Skrift. Norsk Polarinstitut* 160:1–29.

———. 1985. Gradualism and punctuated equilibrium as competing and complementary theories. In J. C. W. Cope and P. W. Skelton, eds., *Evolutionary Case Histories from the Fossil Record. Special Papers in Palaeontology* 33: 17–28.

———. 1988. Seeing is believing: Gradualism and punctuated equilibria in the fossil record. *Science Progress* 72:1–19.

Futuyma, D. J. 1987. On the role of species in anagenesis. *American Naturalist* 130:465–473.

———. 1989a. Speciational trends and the role of species in macroevolution. *American Naturalist* 134:318–321.

———. 1989b. Macroevolutionary consequences of speciation: Inferences from phytophagous insects. In D. Otte and J. A. Endler, eds., *Speciation and Its Consequences,* pp. 557–578. Sunderland, Mass.: Sinauer Associates.

Geary, D. H. 1987. Evolutionary tempo and mode in a sequence of the Upper Cretaceous bivalve *Pleurocardia. Paleobiology* 13:140–151.

———. 1990. Patterns of evolutionary tempo and mode in the radiation of *Melanopsis* (Gastropoda; Melanopsidae). *Paleobiology* 16:492–511.

Gingerich, P. D. 1976. Paleontology and phylogeny: Patterns of evolution of Early Tertiary mammals. *American Journal of Science* 276:1–28.

———. 1983. Rates of evolution: Effects of time and temporal scaling. *Science* 222:159–161.

———. 1984. Punctuated equilibria—where is the evidence? *Systematic Zoology* 33:335–338.

———. 1985. Species in the fossil record: Concepts, trends, and transitions. *Paleobiology* 11:27–41.

Godinot, M. 1985. Evolutionary implications of morphologic changes in Palaeogene primates. In J. C. W. Cope and P. W. Skelton, eds., *Evolutionary Case Histories from the Fossil Record. Special Papers in Palaeontology* 33:39–48.

Goldman, D. and C. E. Mitchell. 1990. Morphology, systematics, and evolution of Middle Devonian Ambocoeliidae (Brachiopoda), western New York. *Journal of Paleontology* 64:79–99.

Goldschmidt, R. 1940. *The Material Basis of Evolution.* New Haven: Yale University Press.

Gould, S. J. 1977. *Ontogeny and Phylogeny.* Cambridge: Harvard University Press.

———. 1980. Is a new and general theory of evolution emerging? *Paleobiology* 6:119–130.

———. 1982. The meaning of punctuated equilibrium and its role in validating a hierarchical approach to macroevolution. In R. Milkman, ed., *Perspectives on Evolution,* pp. 83–104. Sunderland, Mass.: Sinauer Associates.

———. 1986. Punctuated equilibrium at the third stage. *Systematic Zoology* 35:143–148.

———. 1989. Punctuated equilibrium in fact and theory. In A. Somit and S. A. Peterson, eds., *The Dynamics of Evolution,* pp. 54–84. Ithaca: Cornell University Press.

———. 1990. Speciation and sorting as the source of evolutionary trends, or "things are seldom what they seem." In K. J. McNamara, ed., *Evolutionary Trends,* pp. 3–27. London: Belhaven.

———. 1993. The inexorable logic of the punctuational paradigm: Hugo de Vries on species selection. In D. R. Lees and D. Edwards, eds., *Evolutionary Patterns and Processes. Linnean Society Symposium* 14:3–18. London: Academic Press.

Gould, S. J. and N. Eldredge. 1977. Punctuated equilibria: The tempo and mode of evolution reconsidered. *Paleobiology* 3:115–151.

———. 1993. Punctuated equilibrium comes of age. *Nature* 366:223–227.

Grant, P. R. 1986. *Ecology and Evolution of Darwin's Finches.* Princeton: Princeton University Press.

Hallam, A. 1962. The evolution of *Gryphaea. Geological Magazine* 99:571–574.

———. 1982. Patterns of speciation in Jurassic *Gryphaea. Paleobiology* 8:354–366.

Hayami, I. and T. Ozawa. 1975. Evolutionary models of lineage-zones. *Lethaia* 8:1–14.

Henry, J. L. and E. N. K. Clarkson. 1975. Enrollment and coadaptations in some species of the Ordovician trilobite *Placoparia. Fossils and Strata* 4:87–95.

Isaacson, P. E. and D. G. Perry. 1977. Biogeography and morphological conservatism of *Tropidoleptus* (Brachiopoda, Orthida) during the Devonian. *Journal of Paleontology* 51:1108–1122.

Jackson, J. B. C. and A. H. Cheetham. 1990. Evolutionary significance of morphospecies: A test with cheilostome Bryozoa. *Science* 248:579–583.

Johnson, A. L. A. 1985. The rate of evolutionary change in European Jurassic scallops. In J. C. W. Cope and P. W. Skelton, eds., *Evolutionary Case Histories from the Fossil Record. Special Papers in Palaeontology* 33:91–102.

Johnson, G. A. 1982. Occurrence of phyletic gradualism and punctuated equilibria through geologic time. *Journal of Paleontology* 52:1329–1331.

Kat, P. W. and G. M. Davis. 1983. Speciation of molluscs from Turkana Basin. *Nature* 304:660–661.

Kelley, P. H. 1983. The role of within-species differentiation in macroevolution of Chesapeake Group bivalves. *Paleobiology* 9:261–268.

Kellogg, D. E. 1975. The role of phyletic change in *Pseudocubus vema* (Radiolaria). *Paleobiology* 1:359–370.

———. 1983. Phenology of morphologic change in radiolarian lineages from deep-sea cores: Implications for macroevolution. *Paleobiology* 9:355–362.

Kellogg, D. E. and J. D. Hays. 1975. Microevolutionary patterns in Late Cenozoic Radiolaria. *Paleobiology* 1:150–160.

Kirkpatrick, M. 1982. Quantum evolution and punctuated equilibria in continuous genetic characters. *American Naturalist* 119:833–848.

Lande, R. 1976. Natural selection and random genetic drift in phenotypic evolution. *Evolution* 30:314–334.

———. 1980. Genetic variation and phenotypic evolution during allopatric speciation. *American Naturalist* 116:463–479.

Lazarus, D. B. 1983. Speciation in pelagic Protista and its study in the planktonic microfossil record: A review. *Paleobiology* 9:327–340.

———. 1986. Tempo and mode of morphologic evolution near the origin of the radiolarian lineage *Pterocanium prismatium. Paleobiology* 12:175–189.

MacLeod, N. 1991. Punctuated anagenesis and the importance of stratigraphy to paleobiology. *Paleobiology* 17:167–188.

Malmgren, B. A. and W. A. Berggren. 1987. Evolutionary changes in some Late Neogene planktonic foraminiferal lineages and their relationships to paleoceanographic changes. *Paleoceanography* 2:445–456.

Malmgren, B. A., W. A. Berggren, and G. P. Lohmann. 1983. Evidence for punctuated gradualism in the Late Neogene *Globorotalia tumida* lineage of planktonic foraminifera. *Paleobiology* 9:377–389.

Malmgren, B. A. and J. P. Kennett. 1981. Phyletic gradualism in a Late Cenozoic planktonic foraminiferal lineage: DSDP Site 284, southwest Pacific. *Paleobiology* 7:230–240.

Marshall, C. R. 1990. Confidence intervals on stratigraphic ranges. *Paleobiology* 16:1–10.

Mayr, E. 1963. *Animal Species and Evolution*. Cambridge: Harvard University Press.

McKinney, M. L. 1988. *Heterochrony in Evolution: A Multidisciplinary Approach*. New York: Plenum.

McKinney, M. L. and K. J. McNamara. 1990. *Heterochrony: The Evolution of Ontogeny.* New York: Plenum.

McNamara, K. J. 1988. The abundance of heterochrony in the fossil record. In M. L. McKinney, ed., *Heterochrony in Evolution: A Multidisciplinary Approach,* pp. 287–325. New York: Plenum.

———. 1990. *Evolutionary Trends.* London: Belhaven.

Moore, R. C., G. C. Lalicker, and A. G. Fischer. 1952. *Invertebrate Fossils.* New York: McGraw-Hill.

Newman, C. M., J. E. Cohen, and C. Kipnis. 1985. Neo-Darwinian evolution implies punctuated equilibria. *Nature* 315:400–401.

Olson, E. C. and R. L. Miller. 1958. *Morphological Integration.* Chicago: University of Chicago Press.

Otte, D. and J. A. Endler. 1989. *Speciation and its Consequences.* Sunderland, Mass.: Sinauer Associates.

Ozawa, T. 1975. Evolution of *Lepidolina multiseptata* (Permian foraminifera) in East Asia. *Memoir of the Faculty of Science, Kyushu University, Geology* D23:117–164.

Pachut, J. F. 1989. Heritability and intraspecific heterochrony in Ordovician bryozoans from environments differing in diversity. *Journal of Paleontology* 63:182–194.

Pachut, J. F. and R. J. Cuffey. 1991. Clinal variation, intraspecific heterochrony, and microevolution in the Permian bryozoan *Tabulipora carbonaria. Lethaia* 24:165–185.

Palmer, A. R. 1985. Quantum changes in gastropod shell morphology need not reflect speciation. *Evolution* 39:699–705.

———. 1986. Inferring relative levels of genetic variability in fossils: The link between heterozygosity and fluctuating asymmetry. *Paleobiology* 12:1–5.

Palumbi, S. R. 1992. Marine speciation on a small planet. *Trends in Ecology and Evolution* 7:114–118.

Pandolfi, J. M. and C. D. Burke. 1989. Shape analysis of two sympatric coral species: Implications for taxonomy and evolution. *Lethaia* 22:183–193.

Raup. D. M. and R. Crick. 1981. Evolution of single characters in the Jurassic ammonite *Kosmoceras. Paleobiology* 7:200–215.

———. 1982. *Kosmoceras:* Evolutionary jumps and sedimentary breaks. *Paleobiology* 8:90–100.

Raup, D. M. and S. M. Stanley. 1971. *Principles of Paleontology.* San Francisco: W. H. Freeman.

Reyment, R. A. 1975. Analysis of a generic level transition in Cretaceous ammonites. *Evolution* 28:665–676.

———. 1982. Analysis of trans-specific evolution in Cretaceous ostracods. *Paleobiology* 8:293–306.

———. 1985. Phenotypic evolution in a lineage of the Eocene ostracod *Echinocythereis. Paleobiology* 11:174–194.

Robison, R. A. 1975. Species diversity among agnostid trilobites. *Fossils and Strata* 4:219–226.

Ruse, M. 1989. Is the theory of punctuated equilibria a new paradigm? In A. Somit and S. A. Peterson, eds., *The Dynamics of Evolution*, pp. 139–167. Ithaca: Cornell University Press.

Sachs, H. M. and P. F. Hasson. 1979. Comparison of species vs. character description for very high resolution biostratigraphy using cannartid radiolarians. *Journal of Paleontology* 53:1112–1120.

Sadler, P. M. 1981. Sediment accumulation rates and the completeness of stratigraphic sections. *Journal of Geology* 89:569–584.

Schindel, D. E. 1980. Microstratigraphic sampling and the limits of paleontologic resolution. *Paleobiology* 6:408–426.

———. 1982. Resolution analysis: A new approach to the gaps in the fossil record. *Paleobiology* 8:340–353.

Scott, G. H. 1982. Tempo and stratigraphic record of speciation in *Globorotalia puncticulata*. *Journal of Foraminiferal Research* 12:1–12.

Sheldon, P. R. 1987. Parallel gradualistic evolution of Ordovician trilobites. *Nature* 330:561–563.

———. 1993. Making sense of microevolutionary patterns. In D. R. Lees and D. Edwards, eds., *Evolutionary Patterns and Processes. Linnean Society Symposium* 14:19–31. London: Academic Press.

Simpson, G. G. 1961. *Principles of Animal Taxonomy*. New York: Columbia University Press.

Smith, A. B. and C. R. C. Paul. 1985. Variation in the irregular echinoid *Discoides* during the Early Cenomanian. In J. C. W. Cope and P. W. Skelton, eds., *Evolutionary Case Histories from the Fossil Record. Special Papers in Palaeontology* 33:29–38.

Somit, A. and S. A. Peterson. 1989. *The Dynamics of Evolution: The Punctuated Equilibrium Debate in the Natural and Social Sciences*. Ithaca: Cornell University Press.

Sorhannus, U. 1990a. Punctuated morphological change in a Neogene diatom lineage: "local" evolution or migration. *Historical Biology* 3:241–247.

———. 1990b. Tempo and mode of morphological evolution in two Neogene diatom lineages. *Evolutionary Biology* 24:329–370.

Stanley, S. M. 1975. A theory of evolution above the species level. *Proceedings of the National Academy of Sciences, U.S.A.* 72:646–650.

———. 1979. *Macroevolution: Pattern and Process*. San Francisco: W. H. Freeman.

Stanley, S. M. and S. Yang. 1987. Approximate evolutionary stasis for bivalve morphology over millions of years: A multivariate, multilineage study. *Paleobiology* 13:113–139.

Stebbins, G. L. and F. J. Ayala. 1981. Is a new evolutionary synthesis necessary? *Science* 213:967–971.

Tabachnik, R. E. 1992. Identity and individuation of evolutionary entities. *Geological Society of America, Abstracts with Programs* 24:A140.

Templeton, A. R. 1980. Mechanisms of speciation—a population genetic approach. *Annual Review of Ecology and Systematics* 12:23–48.

Thompson, D'A. W. 1942. *On Growth and Form*. Cambridge, U.K.: Cambridge University Press.

Trueman, A. E. 1922. The use of *Gryphaea* in the correlation of the Lower Lias. *Geological Magazine* 59:537–542.

Turner, J. R. G. 1986. The genetics of adaptive radiation: A neo-Darwinian theory of punctuational evolution. In D. M. Raup and D. Jablonski, eds., *Patterns and Processes in the History of Life*, pp. 183–209. Berlin: Springer Verlag.

Van Valen, L. 1969. Variation genetics of extinct animals. *American Naturalist* 103:193–224.

Vrba, E. S. 1980. Evolution, species and fossils: How does life evolve? *South African Journal of Science* 76:61–84.

———. 1985. Environment and evolution: Alternative causes of the temporal distribution of evolutionary events. *South African Journal of Science* 81:229–236.

Vrba, E. S. and S. J. Gould. 1986. The hierarchical expansion of sorting and selection: Sorting and selection cannot be equated. *Paleobiology* 12:217–228.

Waddington, C. H. 1974. *The Evolution of an Evolutionist*. Ithaca: Cornell University Press.

Wei, K. Y. and J. P. Kennett. 1988. Phyletic gradualism and punctuated equilibrium in the Late Neogene planktonic foraminiferal clade *Globoconella*. *Paleobiology* 14:345–363.

West, R. M. 1979. Apparent prolonged evolutionary stasis in the middle Eocene hoofed mammal *Hyopsodus*. *Paleobiology* 5:252–260.

Whatley, R. 1985. Evolution of the ostracods *Bradleya* and *Poseidonamicus* in the deep-sea Cainozoic. In J. C. W. Cope and P. W. Skelton, eds., *Evolutionary Case Histories from the Fossil Record*. *Special Papers in Palaeontology* 33:103–116.

Williamson, P. G. 1981. Palaeontological documentation of speciation in Cenozoic molluscs from the Turkana Basin. *Nature* 293:437–443.

2

Species, Speciation, and the Context of Adaptive Change in Evolution

Niles Eldredge

The theory of punctuated equilibria arose directly from the Modern Synthesis. It represents a welding of a conceptualization of species and speciation developed particularly by Dobzhansky (1937a, 1937b) and Mayr (1942), with an especially bold and creative stance on the relation between evolutionary theory and empirical patterns of the fossil record articulated by Simpson (1944). In applying speciation theory to the interpretation of patterns of phenotypic stasis and change, punctuated equilibria addressed a lingering conflict in the synthesis.

Simpson (see Mayr 1980; Eldredge 1985, 1993) never accepted the "biological species concept," thinking absurd any notion of species that explicitly denies a temporal component to their existence. He firmly believed that species and genera arise in gradual fashion one from another—and that the standard taphonomic account of gaps in the fossil record is sufficient to explain the paucity of good examples of species- and generic-level transitional, transformational change.

Yet, as is well known (see Gould 1980; Eldredge 1985, 1992), Simpson (1944) developed his notion of "quantum evolution" explicitly to address patterns of discontinuity on a larger scale. We might believe that gaps in the record obscure gradual change between species, but higher taxa, Simpson claimed, are another matter. Higher taxa often appear abruptly. Already present in the earliest component species are most of the hallmarks of morphological distinction (i.e., synapomorphies) that forever mark taxa throughout their tens and hundreds of millions of years of subsequent evolutionary history. But, Simpson argued, we cannot invoke the same "gaps in the record" for the origin of higher taxa that we do for species: for if we extrapolate the overall rate of change within, say, bats or whales from the Eocene to the Recent *backward* from their earliest stratigraphic occurrence, we would be forced to conclude that whales and bats arose before any mammals (let alone full-fledged placentals) show up in the fossil record.

Simpson was simply saying that patterns in the fossil record have much to tell us about the evolutionary process. He is especially vivid in the Introduction to *Tempo and Mode in Evolution* where he acknowledges that all the "determinants" of evolution (factors such as mutation rate, population size, generation time, etc.) lie squarely in the realm of genetics. But geneticists can only study "what happens to a hundred rats over the course of ten years." Paleontologists, in contrast, are privy to what happens "to a billion rats in the course of ten million years." Simpson's point actually transcends the usual "consistency argument" (Gould 1980) of the synthesis: recurrent patterns (involving primarily relative rates of phenotypic transformation) set problems for evolutionary theorists to explain using their set of "evolutionary determinants." Simpson's was nothing short of a clarion call for the patterns of evolutionary history to be addressed directly by evolutionary theory. And his theory of quantum evolution was a direct application of his own principles: in quantum evolution, Simpson took on the problem of abrupt discontinuity between taxa of higher categorical rank.

Dobzhansky (1937a) and Mayr (1942) were likewise concerned with discontinuity—but at a finer scale than the patterns addressed in Simpson's quantum evolution. Dobzhansky (1937a, p. 4), echoed directly by Mayr (1942), listed diversity and discontinuity as the twin themes of evolutionary biology. The problem, as he saw it, is that natural selection is both necessary and sufficient to generate a spectrum of (morphological) diversity; but selection alone would be expected to yield a continuous array of morphological diversity. Nature, however, comes in discontinuous packages: "species." According to both Dobzhansky and Mayr, then, something else must be acting to sever the continuous arrays into discrete entities. I (Eldredge 1992, 1993) have elsewhere concluded that Dobzhansky and Mayr adopted a "reproductive community" conceptualization of species precisely because they concluded that reproductive discontinuity must precede general phenotypic discontinuity. They discarded the earlier notion that species are simple collections of similar organisms (who are capable, as a rule, of interbreeding) and replaced it with the idea that species are reproductive communities of organisms (whose similarity devolves from their shared and interchanged genetic information). But species, in Dobzhansky's and Mayr's views, lack discreteness through time—and their picture of gradual transformation of species once established seems much the same as Simpson's.

Thus Simpson saw discontinuity as important at the higher taxonomic levels, while Dobzhansky and Mayr saw it as important primarily at the species level. On the other hand, Simpson thought that classes of patterns involving morphological transformation seen in the fossil record demand explicit attention from evolutionary theorists. In contrast, Dobzhansky and Mayr (neither of whom, after all, was a paleontologist) retained the classical extrapolationist position that what can be gleaned as pattern in the laboratory and field during the lifetime of an investigator must suffice (*faute de mieux*—see Dobzhansky 1937a, p. 12) as a picture of evolutionary pattern whatever the fossil record may happen to look like.

Punctuated equilibria, reduced to its barest essentials, says that the empirical patterns of discontinuity of the sort that Simpson saw as typical of the origin of "higher taxa" actually pertain as well to minimally diagnosable clusters of phenotypes: "species." Stasis—the propensity of species not to exhibit great cumulative change through their stratigraphic ranges (and a phenomenon well known to Darwin's paleontological contemporaries)—merely underscores the distinctness among species-level taxa that is typically present from the earliest recorded occurrence of a species. With Simpson, Gould and I suggested that the pattern arises as a reflection of the evolutionary process—and not as a taphonomic artifact. Unlike Simpson, though, we adopted the Dobzhansky-Mayr conceptualization of species, and attributed the paleontological pattern of species-level discontinuity to speciation. Unlike the Dobzhansky-Mayr version, however, we have noted the empirical reality of stasis—a phenomenon that has increasingly been seen as important (and unanticipated) and demanding close theoretical attention.

Thus punctuated equilibria took the Dobzhansky-Mayr conceptualization of species and, following Simpson's methodological approach, applied it to the fossil record. But punctuated equilibria can also be viewed as a bridge, a resolution to a conflict between the sharply disparate views of neontologists Dobzhansky and Mayr, on the one hand, and paleontologist Simpson on the other. As others have noted (e.g., Gould 1980), there was less in the way of true agreement in the modern synthesis than its proponents often made out to be the case.

The reconciliation of the biological species concept with the apprehension of paleontological patterns of stasis and transformation constitutes merely the core of punctuated equilibria. Of perhaps even greater interest are the two paradoxes implicit in the recognition of such pattern and its explanation in terms of both standard speciation theory—and the more recent idea of species sorting.

Paradoxes in Punctuated Equilibria

If punctuated equilibria reconciled conflicting notions of species and their evolution left unresolved by the architects of the synthesis, it also raised two problems that simply had never arisen under the original Darwinian paradigm. Darwin's original syllogism—that just as a little evolutionary change could accrue over brief intervals of time, large-scale changes would naturally accumulate over proportionally longer intervals of time—remained a core postulate of the synthesis. Directional change under natural selection could simply be extrapolated (e.g., via Simpson's [1944] rubric of "orthoselection") to yield long-term, large-scale evolutionary trends. Moreover, species and higher-level taxa continued to be seen as arbitrarily designated segments of evolving lineages: new taxa arise strictly as a byproduct of the accumulation of adaptive change. Even in the Dobzhansky-Mayr conceptualization of species and speciation, general adaptive evolutionary change is held to be occurring at all times—and is not especially closely associated with speciation per se.

Punctuated equilibria challenges these long-held (if generally unexamined) assumptions, and recognizes instead two paradoxes. Evolutionary change seems not, as the great general rule, to accumulate appreciably during the phyletic history of species, suggesting that morphological change is somehow associated with branching events—that is, with true speciation "events." That itself constitutes the first paradox: given what we know about mechanisms of evolutionary change generally, how is it that adaptive evolutionary change in general could be concentrated phylogenetically around episodes of development of reproductive discontinuity? As I will set forth here, this apparent pattern of association of morphological change with speciation runs counter to both theory and empirical data derived from the modern biota. The goal, clearly, is to resolve the incongruity.

The second paradox follows as an immediate consequence of the first: especially given the prevalence of stasis, how do we explain long-term evolutionary trends, which are recast as a phenomenon of among-species accumulation of directional change? This latter paradox is essentially equivalent to the "paradox of the first tier" (Gould 1985).

The second paradox—where trends under punctuated equilibria are explained with reference to a regime of among-species "sorting" (Vrba 1984a; Vrba and Gould 1986)—has received a great deal more attention than has the first. Beginning with our own discussion (Eldredge and Gould 1972, pp. 111–112 and fig. 5–10), trends have epitomized an entire class of among-species phylogenetic patterns discussed under the rubric of "species selection" (see, inter alia, Stanley 1975 [where he coined the term in this context], 1979; Gould and Eldredge 1977; Eldredge and Cracraft 1980; Vrba 1980 [where the alternative "effect hypothesis" was specified], 1984a; Vrba and Gould 1986; Eldredge 1985, 1989).

Species sorting turns out to be fundamental as well to the analysis of the first paradox, as I will show. But first, I simply note that the term "species selection," and the entire paleobiological debate that has been carried on since 1972 on underlying causes of among-species evolutionary patterns, has caused considerable confusion among the ranks of nonpaleobiological evolutionary biologists. And, it seems, for good reason. Ever since Williams (1966) wrote *Adaptation and Natural Selection* at least in part to counter the group-selectionist views of Wynne-Edwards (1962), evolutionary theorists have been exploring the nature of selection at "levels" other than the purely organismic (or genic—cf. Dawkins 1976). Their overall conclusion seems to be that group selection (at various levels) is a possibility, but cannot have played a significant role in shaping the history of life. How is it that many paleobiologists are convinced that among-species evolutionary patterns cry out for theoretical treatment, while most nonpaleobiological theorists think such phenomena are essentially unimportant?

As is so often the case, the two sides have been talking almost completely at cross-purposes. Paleobiologists have generally not acknowledged that the (slightly older) group selection debate in evolutionary biology *pertains solely to the origin, maintenance, and further modification of organismic phenotypic properties*—in other words, adaptations. In this context, the issue is always: are there group-level phenomena that modify fitness values?

Paleobiologists, in contrast, do not see the issue as purely a matter of differential fitness, or as solely a question of the origin or modification of organismic traits—but rather as a matter of the differential fates of packages of phenotypic properties (i.e., species) within monophyletic clades. Some paleobiologists (e.g., Vrba 1984a; Eldredge 1985) restrict the term "species selection" to true species-level "emergent" properties—as an exact, higher-level analogue of natural selection. In this view, "species sorting" is the general term for among-species evolutionary pattern—as the underlying causality can come from below (as in Vrba's "effect hypothesis"), at the focal level, or even as a reflection of downward causation from a higher level (see especially Vrba 1984a for detailed discussion). Other paleobiologists use the term "species selection" more loosely—essentially, in fact, as a synonym of "species sorting."

But, in any case, no paleobiologist thinks of species selection as an alternative evolutionary mechanism generating adaptive change in the conventional, traditional sense (see, for example, the interchange between Maynard Smith [1987, 1988] and Eldredge and Gould [1988; Gould and Eldredge 1988], where the mutual misunderstanding is especially apparent). Clearly the "group selection" and "species selection" debates are not only separate but quite different. It remains important for paleobiologists to continue to address among-species evolutionary patterns—and, recognizing the source of past confusion in dialogue with other evolutionary biologists, to continue to stress the significance of such patterns for evolutionary theory generally.

Far less attention has been paid to the first paradox: the apparent correlation between adaptive change and speciation events. For although a literal reading of the fossil record proclaims such to be commonly the case, evolutionary biologists conversant with microevolutionary changes in the field find such statements as virtual anathema. Williams (1992) is especially strong in his opposition to this generalization. And it must be conceded that the generalization is counterintuitive. It is time for a much closer look at punctuated equilibria's first paradox.

The First Paradox: Speciation and the Context of Adaptive Stasis and Change

Dobzhansky (1937a) and Mayr (1942) radically transformed the basic concept both of species and, coordinately, of the speciation process in the context of evolutionary theory. (That other species concepts persist is more a reflection of developments in other, albeit related, disciplines—particularly systematics; see Eldredge [1993]). In the Dobzhansky-Mayr view, species are extended reproductive communities. Speciation, then, is a matter of derivation of descendant from ancestral reproductive communities through attainment of "reproductive isolation."

Paterson (e.g., 1985; see Eldredge 1989, 1993) likewise sees species as reproductive communities, defining them as collectivities of organisms sharing a common fertilization system ("Specific Mate Recognition System"—or SMRS). Species, to Paterson, are self-defining real entities in nature. The main difference between the Dobzhansky-Mayr and Paterson views is that Paterson argues against seeing speciation as the development of reproductive isolation between moieties of an ancestrally single species, preferring to see new species arising coincidentally in allopatry as selection acts to insure continued mate recognition within each of the separated parts of the ancestral species. In other words, speciation does not occur (as Dobzhansky [1937a] claimed it did under his "reinforcement" model) as a means by which parts of a differentiated ancestral species can focus their adaptations more precisely on adjacent "adaptive peaks." Selection, Paterson argues, is not for isolation, but simply for ongoing reproductive continuity within isolated populations.

In any case, speciation in modern evolutionary theory is the origin of new reproductive communities from old. All biologists are aware that mother-daughter (or sister) species may remain morphologically virtually indistinguishable ("sibling species") or may differ radically from one another in many traits. Vrba (1980, p. 68, fig. 5; see figure 2.1 here) put the situation extremely well. Both from a priori theoretical principles *and* based on empirical experience with the modern biota, we know that all four possibilities of Vrba's 2 × 2 contingency table (figure 2.1) can be, and in fact *are*, realized. *Within* species, there is a complete spectrum from little differentiation to a great deal of (generally geographically-based) variation. And, as already acknowledged, the same spectrum exists *among* closely related species. It is in this context that the first paradox of punctuated equilibria stands out so vividly: all possibilities can be, and in fact are known to be, realized. Yet the empirical situation in the fossil record (see especially Eldredge 1989, p. 63ff. for review) seems to claim otherwise: morphological change in evolution seems to be heavily concentrated at cladogenetic events—which, under punctuated equilibria, are themselves interpreted as conventional speciation events. This suggests

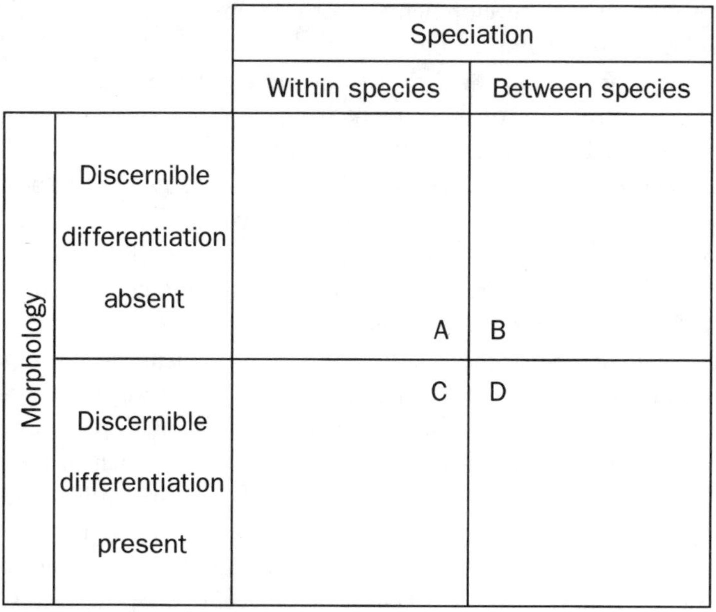

		Speciation	
		Within species	Between species
Morphology	Discernible differentiation absent	A	B
	Discernible differentiation present	C	D

Figure 2.1   Vrba's diagram depicting the "relationship between morphological differentiation and speciation." Redrawn from Vrba 1980, fig. 5, p. 68, as adapted in Eldredge 1989, fig. 4.2, p. 118.

that general adaptive change is correlated with the origin of new reproductive communities—and that, in phylogenetic time, by no means are all four cells in Vrba's 2 × 2 diagram realized with equal probability.

The paradox is even more striking when we recognize the dichotomy between organismic *economic* and *reproductive* features—corresponding to the two great classes of organismic activity that underlie the ecological (economic) and genealogical (reproductive) hierarchies (Eldredge and Salthe 1984; Eldredge 1985, 1986, 1989). Coupled with Paterson's recognition of the SMRS as a species-defining property (see also Lieberman 1992) is the realization that speciation minimally entails change in the SMRS and that it may have nothing whatever to do with transformation of economic phenotypic properties. In other words, SMRS-transformation is both necessary and sufficient for speciation to occur. This contrasts with the older view that sees reproductive isolation primarily as a secondary outgrowth of accumulation of general (i.e., economic) adaptive change in allopatry. In this older model, modification of the fertilization systems is seen primarily as a secondary outcome of the accumulation of sufficient genetic change to inhibit successful reproduction. Paterson's contribution does not invalidate the older model—but simply emphasizes that economic change is not necessary for speciation to occur, as direct modification of any aspect of the SMRS (i.e., including both "pre-" and "post-mating" aspects in the older terminology) may be sufficient for speciation to occur.

Thus reproductive and economic phenotypic properties of sexually reproducing organisms are effectively decoupled, in principle, in evolution. But the phylogenetic outcome, as seen in patterns of within-species stasis, and with economic adaptive change (i.e., as reflected in evolutionary change in the soma, or nonreproductive, aspects of the phenotype) concomitantly correlated strongly with cladogenesis, reveals that economic and reproductive phenotypic properties are by no means decoupled in realized evolutionary history. The problem is to explain the discrepancy between theoretical expectation and empirical reality. In tackling this issue, my plan is, first, to address lingering questions pertaining to the empirical demonstration of species-level patterns of stasis and punctuated change, and the methodological problems inherent in recognizing species in the fossil record. I will then turn to a discussion of the cause(s) of stasis, in the conviction that understanding *why* species do not, as the general rule, demonstrate concerted, wholesale transformation through geologic time has much to tell us about how and why evolutionary change occurs when and as it does. I will conclude with a discussion of factors underlying the correlation of adaptive economic with reproductive evolutionary change—and attempt a reconciliation between theoretical expectation and empirical reality.

Species-Level Patterns in the Fossil Record

Paleobiologists and evolutionary theorists alike have come (at least in large measure) to recognize stasis as an empirically established general phenomenon that had not been anticipated (i.e., predicted) by population genetics theory (e.g., Maynard Smith 1988; Williams 1992). Indeed, Williams (1992) considers stasis one of the most important outstanding problems in evolutionary biology.

Paleobiologists also, as a rule, accept as valid the generalization that species tend not to accumulate much concerted change through their stratigraphic ranges. Debate continues, however, over what constitutes "stasis," and what can fairly be considered "gradual change." One paleontologist's gradualism, it seems, is another's stasis. For example, Eldredge and Gould (1988) question at least three of Sheldon's (1987) eight examples presented as gradual transformation within separate trilobite lineages.

The other side of the punctuated pattern, of course, is the (relatively) rapid and brief spurts of evolutionary transformational change that interrupt much longer periods of stasis. Our original guesstimate that such spurts (interpreted as speciation events) take some 5,000 to 50,000 years (compared with the 5–10 million years of stasis commonly encountered in marine invertebrates) is far from the saltationism imputed to us in some early critiques—and has even been criticized by at least one geneticist (e.g., Lande 1976) for being far too long!

Of perhaps greater interest in the present context is the actual geometry of the punctuated equilibrial pattern itself. Our (Eldredge and Gould 1972) original examples entail overlapping stratigraphic ranges of putative ancestral and descendant species—still the best de facto evidence of true cladogenesis (figure 2.2a). But the fossil record is replete with patterns of the sort depicted in figure 2.2b, where a relatively static species-lineage is succeeded by a presumed descendant lineage with no demonstrable stratigraphic overlap. Such patterns are often interpreted as brief spurts of phyletic transformation, where an entire species (i.e., throughout its entire geographic range) is, in a brief phase of rapid evolutionary transformation, modified into a descendant species—the "punctuated gradualism" of Malmgren et al. (1983).

Vrba (1985) has suggested an alternative explanation of such stepwise patterns—especially if many lineages show the pattern coincidentally in a general episode of faunal turnover (fig. 2.2c). Her "turnover pulse hypothesis" suggests that loss of earlier, and subsequent appearance of later, species is indeed often a matter of biogeographic distributional change reflecting habitat modification. But such habitat changes can also lead to direct extinction of entire species *and to speciation itself*. In other words, habitat alteration leads to (1) emigration and immigration via habitat tracking (an ecological result) and (2) extinction and speciation (an evo-

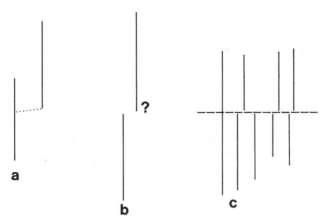

Figure 2.2 Evolutionary patterns in the fossil record. In (a), stratigraphic overlap between putative ancestor and descendant species is the strongest evidence of true speciation. In situations of no overlap (b), several alternative interpretations, including speciation and "punctuated gradualism," are commonly given. The final pattern (c) depicts Vrba's "turnover pulse," involving many species within a single region.

lutionary result). A realistic model can thus explain stepwise patterns within lineages in terms of conventional speciation theory.

It is becoming apparent that much evolutionary change in general (and certainly the evolution of most new species) is tied up with cross-genealogical episodes of extinction and subsequent diversification. Brett and Baird (this volume), for example, document the existence of some 13 time-successive faunas, each with a minimum of 50 (and with a maximum of 335) documented invertebrate species, in a 70 million year interval of the Paleozoic of the Appalachian Basin. Each fauna lasts 5–7 million years, and, on average, there is only some 20% of overlap in species composition in synjacent faunas. Most importantly, from 70% to 85% of the species of each fauna are present in the earliest stratigraphic horizons—and persist to the very end, generally in persistent ecological associations, presenting a pattern that Brett and Baird (this volume) have called "coordinated stasis." Earlier, E. C. Olson (e.g., Olson 1952) documented similar patterns among Permian vertebrate communities. The punctuated pattern, originally described purely in a genealogical context of isolated species-lineages, is emerging as very general indeed.

A NOTE ON FOSSIL SPECIES: Despite the fact that the original examples of punctuated equilibria include well-documented stratigraphic overlap between presumed ancestral and descendant species, Levinton and Simon (1980; see also Levinton 1988) argued that biological species cannot in principle be recognized in the fossil record. To be sure, there are a number of practical difficulties to species recognition with fossils. Most marine invertebrates, for example, reveal little or nothing of the SMRS in their exoskeletons. (The SMRS of sessile benthic organisms—except, notably, barnacles—is perforce restricted to chemical compatibilities of eggs and sperm and to shared signals and schedules for release of same in sea water.) Moreover, both sibling species and polymorphism will inevitably cause under- and overestimation of species, respectively.

Similar problems affect biologists working with the modern biota, where reproductive data are often lacking. As is well known (see Mayr 1942, pp. 148 ff. for extensive discussion), local populations of different species living in the same area are usually quite discrete. Problems in determining species identity and composition, in other words, are far more likely to arise when comparing allopatric populations than when assessing the biota locally. The same is obviously true of the fossil record—and the prevalence of stasis just makes the recognition of such discrete packages easier than would otherwise be expected.

In principle we should expect to find spatiotemporally localized fossil biotas partitioned into more or less discrete phenotypic entities. This is precisely what paleontologists do encounter on bedding planes, and the reasonable inference is that each cluster is a sample of a local population of a discrete reproductive community—a species. Moreover, those discrete clusters are usually recognizable at other stratigraphic horizons. Despite the double pitfalls posed by sibling species and polymorphism (leading to the aforementioned under- and overestimation of species present in a sample), the task of species recognition in paleontology is inherently neither more nor less tricky a proposition than it is with the Recent biota.

We know, empirically, that species have births (speciation) and deaths (extinction). Along the line between birth and death (and I will argue that the temporal interval between birth and death is often short), the species undergoes a history that may (or may not) involve accumulation of significant amounts of (genetically based) phenotypic modification. This is the empirical value of Ghiselin's suggestion that species are "individuals": species are real entities in nature (Ghiselin 1974; see also Hull 1976). And if species are spatiotemporally bounded historical entities (see also Wiley 1978), it is our obligation, as paleontologists, to identify them in the fossil record—taking note of the pitfalls that may attend the recognition of species in any given instance.

Toward an Explanation of Stasis

We are now in a position to ask why economic adaptive change seems to be highly correlated with speciation (reproductive change) in phylogenetic history. I approach the problem by first asking why economic change is *not* commonly expressed in the form of wholesale transformation of organismic phenotypic properties within an entire species. In other words, what underlies stasis?

Darwin (1859) had argued so persuasively for evolution through natural selection that the common perception that evolution could not happen was instantly transformed to the maxim that evolution is inevitable given the mere passage of time. Simpson (1953) echoed these sentiments when he said that it is difficult to see how evolution *cannot* occur. More recently, Williams (1992) has come to see stasis as itself an important evolutionary paradox: on the one hand, species lineages (he agrees) seem remarkably stable; on the other, though, adaptive diversification seems rampant.

Williams (1992), interestingly, resorts to a form of species selectionism, which he terms "normalizing clade selection," as a model to explain stasis. He sees anatomically and ecologically specialized lineages as comparatively short-lived, subject to constant pruning back—while the more generalized (primitive, plesiomorphic, "ancestral") forms live on. His model explicitly addresses among-species, within-clade patterns, such as Bell's stickleback data (e.g., Bell 1989). Virtually identical patterns, where plesiomorphic lineages of rather few species form the sister group of a much more speciose and rapidly evolving lineage, have been reported by many paleobiologists in recent years (most notably, Vrba's 1980 and 1984b alcelaphine-aepycerotine antelope data, as well as many other examples in Eldredge and Stanley 1984).

Though, as already mentioned, evolutionary theorists have conceded that they failed to anticipate the phenomenon of stasis in their evolutionary models, ironically the expectation of stasis is fully implicit in Wright's "shifting balance theory." Wright, though a theoretician and statistical analyst of experimental data (and *not* a field biologist), more than anyone else is responsible for our received views on the organization of species in nature (Wright 1931, 1932; for discussion see Eldredge 1985). He saw species as broken up into quasi-independent "colonies" (later "demes")— a view embraced by Dobzhansky (1937a), himself the epitome of a field-oriented evolutionary geneticist.

Wright's shifting balance theory addresses the evolutionary fate of alleles within species partitioned into quasi-isolated colonies. Under certain conditions, an allele can indeed spread rapidly throughout an entire species—and the much discussed rapid spread of the p element in *Drosophila melanogaster* (Kidwell, Novy, and Feely 1981) seems to bear out the plausibility of such events. But consider the more general result: under shifting balance, each deme (local population) will be expected to undergo slightly different microevolutionary histories. Each local population, integrated into its own local ecosystem, will start with its own sampling of the total species genetic variance, will experience different mutational phenomena, and will of course be under somewhat different selective regimes. Some populations become extinct, while others merge with neighboring populations.

Wright's view of species organization, then, makes it easy to see how differentiation can occur within species through both natural selection and genetic drift. But it also implies the unlikelihood that species-wide change will accrue in lockstep fashion through geologic time. Indeed, as I have pointed out elsewhere (Eldredge 1989, p. 87), under shifting balance, we would expect most cases of gradual directional change in the fossil record to be confined to restricted subsets of single species. And such indeed appears to be the case: Gingerich's well-known examples of putative gradual change within several Eocene mammalian lineages (e.g., Gingerich 1976) are restricted to the Bighorn Basin—where by no means are these lineages endemic. The eight trilobite lineages in Sheldon 1987 come from the Builth Inlier, almost certainly not the entire range of any of the lineages involved. And, because both sets of data can be interpreted more as oscillatory stasis than classic directional gradualism, I cite an example from my own early work on the trilobite *Phacops rana*. I (Eldredge 1972) documented an increase of overall number of lenses in relation to standardized cephalon size over an interval of several million years within this species—but the trend was only found in samples from central and western New York State, a small subset of the entire geographic range of this species. Interestingly, the pattern of geographic differentiation in this feature was maintained throughout the duration of the trend.

The real key to stasis, I suspect, lies in the realization that the supposition that evolutionary change is inevitable given the mere passage of time is itself based on a flawed assumption: It was known in the nineteenth century that environments are always subject to change in geological time. Evolutionary change seemed inevitable because natural selection (given the requisite heritable variation) was seen as the means whereby species (meaning phenotypic properties of organisms within species) would become modified to track environmental change.

But evolutionary change is not what generally happens when environments change. As we (Eldredge and Gould 1974) argued, there are three possible outcomes of environmental change that eliminate the capacity of organisms of any given species to persist at any one place. The most likely outcome is simple biogeographic change. Species ranges are continuously revamped as a reflection of environmental change. Many mammals, birds, and plants are well documented as moving north within North America during the twentieth century (see, for example, Davis et al. 1986). More graphically, Pleistocene histories of innumerable species, both marine and terrestrial, are famous for their accordion-like latitudinal swings as climate alternately cools and warms. The next most likely option is total extinction—not just of local populations but of an entire species. The least likely alternative, it seems, is *in situ* evolutionary change to meet the new conditions.

Thus, coupled with expectations implicit in Wright's work, habitat tracking appears to be an important ingredient of stasis. As long as organisms can recognize suitable habitat, they will tend to persist, and to do so unchanged. Indeed, Lieberman et al. (in press) make the intriguing suggestion that habitat tracking is subordinate to the simple fact that evolution cannot possibly proceed in a lockstep, phyletic fashion within a species that is diversified into numerous different habitats and ecosystems. They base their conclusions on detailed analysis of evolutionary patterns within two species of Middle Devonian (Hamilton) brachiopod species. Whether or not the sorts of patterns they document are general, stasis is the overwhelming rule in virtually all of the 300 + species of the Middle Devonian Hamilton Group—despite the constant spatial shifting of entire communities (themselves remarkably compositionally stable) throughout the 5–6 million year interval (Brett and Baird in this volume).

Habitat tracking—stabilizing natural selection in the face of environmental change—is vital for explaining the phenomenon of stasis.

Williams (1992, p. 130) dismisses habitat tracking as a cause of stasis, pronouncing it a "fable." He then proceeds to elaborate a "desperation hypothesis," where stasis emerges as a statistical result of the differential elimination of divergent subclades. Citing Bell's admirable studies of fossil and Recent three spine sticklebacks (e.g., Bell 1989), Williams argues that specialized, freshwater populations are selectively eliminated—basically because their habitats are ephemeral. The marine, ancestral form lives on. Why? Williams writes (1992, pp. 134–135):

> Most species [i.e., of sticklebacks] would be constantly breaking up into divergent lines specializing for new habitats or ways of life. Then after a few millennia most such lines have gone extinct. The remaining few nearly always include a form close to the ancestral type. If this form were numerically dominant and most widespread, like the marine stickleback, it would be most likely to contribute to the fossil record and appearance of stasis.
>
> The usual persistence of the ancestral type would follow from purely statistical considerations. Not all habitats or ways of life need be equally durable. If we had to predict which populations will still be represented by descendants a million years from now, we would do well to choose those that have already persisted with little change for millions of years, rather than their recent offshoots of mere thousands or tens of thousands of years in age. The appearance of stasis in the fossil record would result from an enormous variability in the persistence of ecological niches.

What is this argument but an appeal to persistence of habitat to explain persistence of relatively unchanged morphology—in other words, the very "fable" of habitat tracking Williams so cavalierly rejects a few pages earlier?

As to Williams's suggestion that we do not really see all the side-branch spurts of evolutionary change (of the sort so well displayed in Bell's stickleback data), he might be right that the pattern is more general. We do tend to see in the fossil record mostly those species that are abundant and widespread: there *is* a bias against the preservation of small populations or narrowly distributed species, which might very well be more likely to show divergent evolutionary change. Even so, it is the persistence of the morphology that we *do* see that is the core issue of stasis—and Williams offers no explicit explanation of such persistence that differs in any discernible way from habitat tracking: the persistence of habitats (he prefers "niches") in the face of inevitable environmental change over geologic time. For if ever there were a myth, it is that the environment of any particular locale can remain stable over long periods of geologic time.

Sheldon (1993) makes a related point by arguing that stasis should be more prevalent in eurytopic lineages than in stenotopes. He links his argument directly to notions of environmental stability: phyletic gradualism would be the expected mode in relatively stable environments (e.g., offshore marine waters), while stasis, ironically, might be more prevalent in fluctuating environments. Elsewhere (e.g., Eldredge 1989) I have utilized (and will utilize here) niche-width arguments to similar purpose—though my emphasis is on differential probability of survival of taxa of varying niche widths. Completion of current detailed research on stasis of the 300 + species of Hamilton Group invertebrates in the context of ecological community distribution will provide a direct test of Sheldon's suggestion. Experience so far seems to indicate that stasis is equally prevalent within nearshore and offshore species.

Stasis, then, is to be expected even in times of environmental change. We have come nearly full circle. Darwin took us from "evolution is impossible" to "evolution is inevitable." We now see that evolutionary change is indeed easily effected by natural selection—but that its accumulation in geologic time is far from the simple inevitability we have so long assumed it to be. Evolutionary change seems largely evanescent, and likely to accumulate in a meaningful, phylogenetic sense only under certain conditions and circumstances—which I shall now attempt to specify.

The Evolutionary Context of Adaptive Change

Wright's shifting balance theory pictures a species broken up into a number of quasi-isolated, semi-evolutionarily independent local populations. I have argued that Wright's model of the internal structure and evolutionary dynamics of species actually predicts species-wide stasis. Returning to the Dobzhansky-Mayr stricture that reproductive isolation imposes discontinuity on an otherwise continuous array of adaptive differentiation, we can begin to see why total phylogenetic independence is necessary in order for adaptive differentiation to be literally injected into the phylogenetic stream: within-species differentiation can proceed only so far and is destined to be lost or at least recombined if the differentiation remains wholly within a single, albeit somewhat disaggregated, reproductive community.

Speciation—attainment of "reproductive isolation" in Dobzhansky-Mayr parlance—partitions genetic variance. Futuyma (1987) makes this very point in attempting to explain punctuated patterns of speciation and adaptive change. Speciation acts like a ratchet, lopping off bits of within-species, among-deme variation, thus creating an independent phylogenetic entity with its own "unitary evolutionary role and tendencies" (to crib a bit of Simpson's definition of "evolutionary species," [e.g., Simpson 1961, p. 153]).

We (Eldredge and Gould 1974) explicitly addressed the paradox of the empirical association of adaptive economic change with true speciation, based on the prevailing model of allopatric speciation that stresses the role of peripheral isolates in the speciation process. Punctuated equilibria has been criticized as overly tied to one particular model of speciation. While we freely grant that many such models exist and that most (perhaps even all) of them are good descriptors of what may well happen at times in nature, there is perhaps more reason now than in the early 1970s to take the peripheral isolate model of speciation seriously.

Stevens (1989, 1992) has discussed the role and importance of the ecological "living dead." All species have (at any one moment) finite geographic ranges. The ranges are typically the limits of (reachable) recognizable habitat. The "living dead" are the organismic outposts at the ecologic periphery of species ranges—where edaphic conditions are at the extremes of tolerance limits, and populations are typically, as a consequence, sparse and often consist of organisms of reduced "vigor." Reproduction is greatly hindered under such circumstances.

We (Eldredge and Gould 1974) simply supposed that, given the onset of reproductive isolation for a population at the periphery of a species's range, and given requisite heritable variation, selection might well rather quickly "redefine" as optimal a habitat that had been marginal for the ancestral species. Thus adaptive change is correlated with speciation (1) because of the very partitioning action of reproductive isolation, and (2) because, under some conditions at least, adaptive change is expected to be triggered by such isolation.

Peripheral isolate speciation offers one such set of conditions—where it can be supposed that the process works with little or no change in the habitats per se. The speciation component of Vrba's (1985) "turnover pulse hypothesis," in which species fragmentation is directly associated with habitat change, is clearly another set of circumstances favoring the association of economic adaptive change with actual speciation—that is, reproductive adaptive change.

Species Sorting and the Generation of Evolutionary Pattern

Peripheral isolate speciation and the turnover pulse hypothesis offer causal explanations of why adaptive change generally seems to be associated with speciation. But there is another component to the evolutionary process that has empirical ramifications for the apprehension of pattern in the fossil record. Critics of punctuated equilibria, especially in the late 1970s, took our postulate to mean that evolutionary change in general is saltatory, and that prodigious amounts of change always accompany speciation events. This misreading lingers, despite our original examples (and subsequent discussions) of punctuated change displaying characteristically only modest amounts of anatomical change.

One purely methodological ramification is that we can only see those species that are in fact well differentiated from their parental species—the issue of sibling species discussed earlier. If sibling species are equally common as well-differentiated species, then it is not true that speciation must always entail significant economic adaptive change—though it remains the case that when such significant differentiation does occur, it is associated with cladogenesis. In other words, the pattern may in part be artifactual, beclouded by the existence of undetected sibling species.

I propose a slightly different explanation, one that remains true to the spirit of Simpson's view of the fossil record as more a reflection of evolutionary process than of artifact. Evolutionary biologists are, as we have seen, in essential agreement that, while sibling species are not all that common, nonetheless most speciation events entail relatively minor adaptive change. Then why do we seem to see a marked bias toward well-differentiated species-level taxa in the fossil record?

Let us return to the supposition, made stronger by Paterson's (1985) views on species and speciation, that economic and reproductive adaptive change are, at least in principle, decoupled. Lewis (1966) reviewed rapid speciation events in flowering plants. He found that new species can arise through rapid chromosomal differentiation leading to hybrid sterility. But such new species are commonly ecologically undifferentiated from the parental species. Unable to gain an ecological foothold, the fate of nearly all such fledgling species is extinction.

Lewis's discussion suggests a very general model—one in which species sorting plays a crucial role, acting as a ratchet mechanism in the phylogenetic accumulation (*not* the initial generation) of adaptive change. The model asserts that the speciation process in general produces more relatively undifferentiated descendant species than descendants markedly divergent from their parental species. *But extinction probability is inversely related to degree of ecological, adaptive differentiation.* The more ecologically divergent the descendant, according to this model, the greater the chances it will survive, that it will realize those "unitary evolutionary roles and tendencies"—and that it will show up for us to find them in the fossil record. Species sorting, hinging on the survival probability of fledgling species, provides a ratchetlike mechanism for the accumulation (*not* the generation or origination) of adaptive change through time.

The present model of ratcheting adaptive change through species sorting recalls earlier discussions (Eldredge 1979, 1985, 1989; Eldredge and Cracraft 1980; Vrba 1980, 1984b) relating speciation and extinction rates, and their involvement with the generation of macroevolutionary pattern, explicitly to parameters of niche width. These models, like Williams's (1992) concept of clade selection, pertain to differentials in both speciation rate and in *long-term* species survival. Specifically, I have argued (in references just cited) that species composed of relatively eurytopic organisms characteristically have lower speciation *and* extinction rates when compared to sister taxa in which stenotopy prevails. Such stenotopic lineages are characteristically far more speciose (at any one time) than are eurytopic sister taxa.

A combined model, then, sees speciation (SMRS change) as perhaps equally probable in both eurytopes and stenotopes, but stenotopes have a higher rate of survival because they tend to show ecological specialization, hence differentiation, more than eurytopic lineages do (as in the above ratcheting model). But, once established, species whose organisms are relatively eurytopic tend to persist longer than specialized stenotopes, which are far more prone to extinction. Relatively high ecological differentiation lowers extinction probability in fledgling species, but over the long haul leads to higher extinction rates than among eurytopes—which tend to be ecologically less differentiated on an among-species, within-lineage basis.

Mass extinctions and their evolutionary aftermath offer an additional pattern to which this model may well apply. If speciation is primarily a matter of formation of new reproductive communities and is a causal context, rather than an effect, of the accumulation of adaptive change, how can we continue to regard rapid proliferation of species following episodes of extinction as driven by the existence of a large number of now-vacated ecological niches?

Vrba's (1985) turnover pulse hypothesis, which postulates that the very same phenomena of habitat alteration and disruption that underlie the extinction of some species may well be the impetus for speciation in other lineages, goes far toward providing a causal framework for understanding species proliferation following extinction: speciation rate goes up along with extinction rate. But most of the truly great extinctions have distinct, nontrivial temporal lags between massive disappearance of species and the subsequent proliferation of new taxa. Such lags would seem to demand additional explanation beyond the turnover pulse hypothesis.

I suggest that, rather than seeing speciation rate per se increase as a consequence of massive species loss, the probability of survival of fledgling species simply increases. Much the same thing happens, after all, as population growth resumes following a crash: birth rate does not increase so much as the survival of progeny increases until population size reaches equilibrial limits. As with organisms within species, why not with species within monophyletic higher taxa?

If this general line of reasoning is on the right track, species sorting emerges as an even more important key to understanding the geometry of historical evolutionary pattern than has previously been supposed. Nor has my discussion been exhaustive: I have assumed that the probability of SMRS disruption is equal in sister taxa—and have equated species sorting with differential probability of extinction. Such need not be the case. Vrba (1980, 1984a) long ago pointed out that sorting can arise through differential probability of species origination (SMRS disruption), and Lieberman, Allmon, and Eldredge (1993) present an example involving developmental mechanisms that may in fact induce such a bias. Much more work is to be done on species sorting. But by now it is clear that species sorting provides the context in which the fate of adaptive differentiation via natural selection is largely determined.

Acknowledgments   I thank Bruce Lieberman, Ian Tattersall, Elisabeth S. Vrba, and the reviewers for their helpful comments on earlier drafts of this paper.

References

Bell, M. A. 1989. Stickleback fishes: Bridging the gap between population biology and paleobiology. *Trends in Ecology and Evolution* 3:320–325.

Darwin, C. 1859. *On the Origin of Species*. London: John Murray.

Davis, M. B., K. D. Woods, S. L. Webb, and R. P. Futuyma. 1986. Dispersal versus climate: Expansion of *Fagus* and *Tsuga* into the Upper Great Lakes region. *Vegetatio* 67:93–103.

Dawkins, R. 1976. *The Selfish Gene*. New York and Oxford: Oxford University Press.

Dobzhansky, T. 1937a. *Genetics and the Origin of Species*. Reprint edition, 1982. New York: Columbia University Press.

———. 1937b. Genetic nature of species differences. *American Naturalist* 71:404–420.

Eldredge, N. 1972. Systematics and evolution of *Phacops rana* (Green, 1832) and *Phacops iowensis* Delo, 1935 (Trilobita) from the Middle Devonian of North America. *Bulletin of the American Museum of Natural History* 147:45–114.

———. 1979. Alternative approaches to evolutionary theory. In J. H. Schwartz and H. B. Rollins, eds., *Models and Methodologies in Evolutionary Theory. Bulletin of the Carnegie Museum of Natural History* 13:7–19.

———. 1985. *Unfinished Synthesis: Biological Hierarchies and Modern Evolutionary Thought*. New York: Oxford University Press.

———. 1986. Information, economics and evolution. *Annual Review of Ecology and Systematics* 17:351–369.

———. 1989. *Macroevolutionary Dynamics: Species, Niches and Adaptive Peaks*. New York: McGraw-Hill.

———. 1992. Marjorie Grene, *Two Evolutionary Theories*, and modern evolutionary theory. *Synthese* 92:135–149.

———. 1993. What, if anything, is a species? In B. Kimbel and L. Martin, eds., *Species, Species Concepts and Primate Evolution*, pp. 3–20. New York: Plenum.

Eldredge, N. and J. Cracraft. 1980. *Phylogenetic Patterns and the Evolutionary Process: Method and Theory in Comparative Biology.* New York: Columbia University Press.

Eldredge, N. and S. J. Gould. 1972. Punctuated equilibria: An alternative to phyletic gradualism. In T. J. M. Schopf, ed., *Models in Paleobiology,* pp. 82–115. San Francisco: Freeman, Cooper.

———. 1974. Reply to Hecht. *Evolutionary Biology* 7:303–308.

———. 1988. Punctuated equilibrium prevails. *Nature* 332:211–212.

Eldredge, N. and S. N. Salthe. 1984. Hierarchy and evolution. *Oxford Surveys in Evolutionary Biology* 1:182–206.

Eldredge, N. and S. M. Stanley, eds. 1984. *Living Fossils.* New York: Springer Verlag.

Futuyma, D. J. 1987. On the role of species in anagenesis. *American Naturalist* 130:465–473.

Ghiselin, M. T. 1974. A radical solution to the species problem. *Systematic Zoology* 23:536–544.

Gingerich, P. D. 1976. Paleontology and phylogeny: Patterns of evolution at the species level in Early Tertiary mammals. *American Journal of Science* 276: 1–28.

Gould, S. J. 1980. G. G. Simpson, paleontology, and the modern synthesis. In E. Mayr and W. B. Provine, eds., *The Evolutionary Synthesis: Perspectives on the Unification of Biology,* pp. 153–172. Cambridge: Harvard University Press.

———. 1985. The paradox of the first tier: An agenda for paleobiology. *Paleobiology* 11:2–12.

Gould, S. J. and N. Eldredge. 1977. Punctuated equilibria: The tempo and mode of evolution reconsidered. *Paleobiology* 3:115–151.

———. 1988. Species selection: Its range and power. *Nature* 334:19.

Hull, D. J. 1976. Are species really individuals? *Systematic Zoology* 25:174–191.

Kidwell, M. G., J. B. Novy, and S. M. Feely. 1981. Rapid unidirectional change of hybrid dysgenesis potential in *Drosophila. Journal of Heredity* 72:32–38.

Lande, R. 1976. Natural selection and random genetic drift in phenotypic evolution. *Evolution* 30:314–334.

Levinton, J. 1988. *Genetics, Paleontology and Macroevolution.* Cambridge, U.K., and New York: Cambridge University Press.

Levinton, J. S. and C. M. Simon. 1980. A critique of the punctuated equilibria model and implications for the detection of speciation in the fossil record. *Systematic Zoology* 29:130–142.

Lewis, H. 1966. Speciation in flowering plants. *Science* 152:167–172.

Lieberman, B. S. 1992. An extension of the SMRS concept into a phylogenetic context. *Evolutionary Theory* 10:157–161.

Lieberman, B. S., W. D. Allmon, and N. Eldredge. 1993. Levels of selection and macroevolutionary patterns in the turritellid gastropods. *Paleobiology* 19:205–215.

Lieberman, B. S., C. E. Brett, and N. Eldredge. 1995 (in press). Patterns and processes of stasis in two species lineages from the Middle Devonian of New York State. *Paleobiology* 21.

Malmgren, B. A., W. Berggren, and G. P. Lohmann. 1983. Evidence for punctuated gradualism in the Late Neogene *Globorotalia tumida* lineage of planktonic foraminifera. *Paleobiology* 9:377–389.

Maynard Smith, J. 1987. Darwinism stays unpunctured. *Nature* 330:516.

———. 1988. Punctuation in perspective. *Nature* 332:311–312.

Mayr, E. 1942. *Systematics and the Origin of Species*. Reprint edition, 1982. New York: Columbia University Press.

———. 1980. Biographical essays: G. G. Simpson. In E. Mayr and W. B. Provine, eds., *The Evolutionary Synthesis*, pp. 452–463. Cambridge: Harvard University Press.

Olson, E. C. 1952. The evolution of a Permian vertebrate chronofauna. *Evolution* 8:181–196.

Paterson, H. E. H. 1985. The recognition concept of species. In E. S. Vrba, ed., *Species and Speciation. Transvaal Museum Monograph* 4:21–29.

Sheldon, P. R. 1987. Parallel gradualistic evolution of Ordovician trilobites. *Nature* 330:561–563.

———. 1993. Making sense of microevolutionary patterns. In D. R. Lees and D. Edwards, eds., *Evolutionary Patterns and Processes. Linnean Society Symposium* 14:19–31. London: Academic Press.

Simpson, G. G. 1944. *Tempo and Mode in Evolution*. New York: Columbia University Press.

———. 1953. *The Major Features of Evolution*. New York: Columbia University Press.

———. 1961. *Principles of Animal Taxonomy*. New York: Columbia University Press.

Stanley, S. M. 1975. A theory of evolution above the species level. *Proceedings of the National Academy of Sciences* 72:646–650.

———. 1979. *Macroevolution: Pattern and Process*. San Francisco: W. H. Freeman.

Stevens, G. C. 1989. The latitudinal gradient in geographical range: How so many species coexist in the tropics. *American Naturalist* 128:35–46.

———. 1992. Spilling over the competitive limits to species coexistence. In N. Eldredge, ed., *Systematics, Ecology, and the Biodiversity Crisis*, pp. 40–58. New York: Columbia University Press.

Vrba, E. S. 1980. Evolution, species and fossils: How does life evolve? *South African Journal of Science* 76:61–84.

———. 1984a. What is species selection? *Systematic Zoology* 33:318–328.

———. 1984b. Evolutionary pattern and process in the sister-group Alcelaphini-Aepycerotini (Mammalia: Bovidae). In N. Eldredge and S. M. Stanley, eds., *Living Fossils*, pp. 62–79. New York: Springer-Verlag.

————. 1985. Environment and evolution: Alternative causes of the temporal distribution of evolutionary events. *South African Journal of Science* 81:229–236.

Vrba, E. S. and S. J. Gould. 1986. The hierarchical expansion of sorting and selection: Sorting and selection cannot be equated. *Paleobiology* 12:217–228.

Wiley, E. O. 1978. The evolutionary species concept reconsidered. *Systematic Zoology* 27:17–26.

Williams, G. C. 1966. *Adaptation and Natural Selection: A Critique of Some Current Evolutionary Thought*. Princeton: Princeton University Press.

————. 1992. *Natural Selection: Domains, Levels, and Applications*. New York: Oxford University Press.

Wright, S. 1931. Evolution in Mendelian populations. *Genetics* 16:97–159.

————. 1932. The roles of mutation, inbreeding, crossbreeding, and selection in evolution. *Proceedings of the Sixth International Congress of Genetics* 1:356–366.

Wynne-Edwards, V. C. 1962. *Animal Dispersion in Relation to Social Behavior*. Edinburgh: Oliver and Boyd.

II

Speciation Patterns and Processes

3

The Importance of Gradual Change in Species-Level Transitions

Dana H. Geary

The study of tempo and mode in the fossil record bears on a number of fundamental issues in evolutionary biology, in particular, the nature of species and species-level change. After twenty years of empirical investigation into the tempo and mode of species-level change in the fossil record, it is clear that both punctuated equilibria and gradualism occur, as do a variety of intermediate patterns.

The theory of punctuated equilibria generated intense scrutiny but has largely borne it well. Many well-documented fossil sequences exhibit geologically instantaneous change (e.g., Cheetham 1986), and extended intervals of no net morphological change have proven to be quite common (e.g., Stanley and Yang 1987). Geologically rapid change is understood to encompass tens of thousands of years in most cases, certainly long enough for natural selection to bring about significant differences.

The focus of this paper is on tempos that are *not* geologically instantaneous. A number of such patterns have been described; I argue here that the importance of these patterns for understanding species and species-level changes has not been fully explored.

Evolutionary Tempo: Examples from Molluscs

A wide range of evolutionary tempos has been observed in the fossil record (e.g., Malmgren et al. 1983; Bell et al. 1985; Cheetham 1986; Lazarus

1986; Stanley and Yang 1987; Sheldon 1987; Wei and Kennett 1988; Sorhannus 1990). Examples from my own work and that of my students illustrate some of this range.

Melanopsid Gastropods

Melanopsid gastropods (Family Melanopsidae, Superfamily Cerithioida) from the Pannonian Basin system in eastern Europe exhibit a variety of evolutionary tempos (Geary 1990a,b, 1992; Geary et al. 1989). Two lineages of melanopsids are present in Middle and Late Miocene deposits, shown in figure 3.1 as the *Melanopsis bouei* clade (at right) and the *M. impressa* clade (at left). (Recent work by Pal Müller [1992] of Budapest suggests that these two lineages were distinct since at least the Oligocene and that generic level distinction between them is probably warranted.) A burst of new species occurs in the *M. bouei* clade in the middle part of the Pannonian Stage. Several of these species are quite common in Pannonian Basin deposits (e.g., *M. pygmaea, M. inermis*) yet their first appearances are abrupt, and preceded by no intermediate forms. Temporal

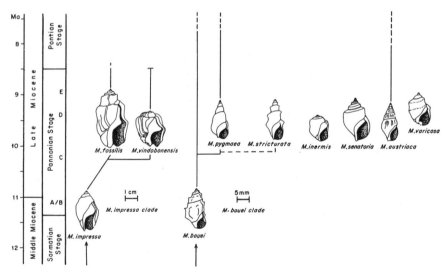

Figure 3.1   Two clades of melanopsid gastropods from the Pannonian Stage of the Pannonian Basin. Note different scale bars for each clade. Source: Geary 1990.

resolution here is not particularly good; the intervals during which these species evolved may have been many tens of thousands of years.

Morphological patterns in the *M. impressa*—*M. fossilis* lineage provide a striking contrast. During the Middle Miocene, *M. impressa* underwent an interval of stasis lasting at least seven million years. This was followed by directional increases in shell size (figure 3.2a) and shouldering (figure 3.2b) that occurred over a period of roughly two million years (Geary 1990b). The resulting morphology is strikingly different from that of the ancestral species. The Pannonian Basin was several hundred kilometers wide and topographically complex, and localities are widely separated around the basin. Nonetheless, surprisingly little evidence of geographic variation exists in the *M. impressa*—*M. fossilis*—*M. vindobonensis* lineage.

In the latest Miocene Pontian Stage, the *M. caryota*—*M. cylindrica* lineage (descendants of *M. impressa*) exhibits a gradual increase in shell shouldering (figure 3.3; Staley 1992). Temporal resolution of this sequence is not very good, but the interval of change is once again long, probably between 500,000 and one million years. The iterative nature of this change is the subject of Staley et al. (submitted).

In what is perhaps the most unusual temporal pattern, the Pannonian species *M. fossilis* and *M. vindobonensis* (an ancestor-descendant pair) occur together with morphological intermediates for an interval lasting approximately one million years beyond their first appearances (Geary 1992). Each histogram in figure 3.4 records the scores of individuals along a discriminant axis, chosen to best distinguish typical *M. vindobonensis* (at right) from typical *M. fossilis* (at left). Intermediates occur in all samples in which both forms are present. Following this extended interval of what I interpret as incomplete reproductive isolation, the two species become distinct at the end of the Pannonian Stage, as shown in the two youngest samples.

In summary, melanopsid gastropods exhibit a range of evolutionary tempos. Some sequences are abrupt and without intermediates. Others are more continuous transitions that occur with a full range of intermediate morphologies, lasting for up to two million years. Finally, in the example just described, distinctive morphologies arise quickly, but reproductive isolation between them occurs more slowly. As Fortey (1985) pointed out, recognition of disparate tempos from within the same sections reinforces our overall confidence in these morphological patterns.

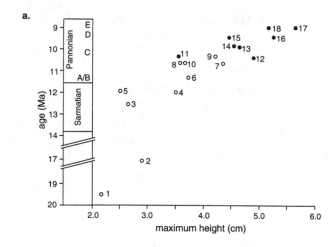

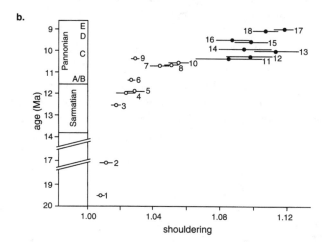

Figure 3.2 Maximum height (a) and shouldering (b) versus age for Sarmatian and Pannonian Stage samples of *Melanopsis impressa* (open circles) and *M. fossilis* (solid circles). Relative temporal positions within the following sets of samples are uncertain: (4–5), (7–10), (11–12), (15–16), (17–18). Error bars in (b) represent +/− 2 standard errors about the mean. Shouldering is measured as the ratio of two lengths (the length along the shell perimeter from the apex to the point of maximum width, divided by the straight-line distance connecting the same two points). After Geary 1990.

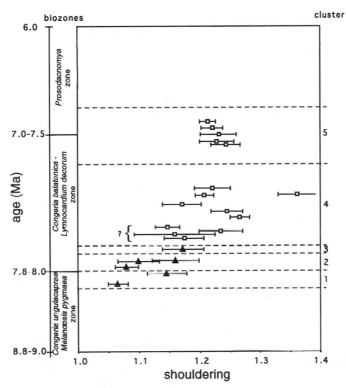

Figure 3.3 Shouldering versus approximate age for Pontian Stage samples of *Melanopsis caryota* (triangles) and *M. cylindrica* (squares). Clusters represent subdivisions of the local biostratigraphy. Temporal position of the bracketed samples in cluster 4, relative to one another and to other cluster 4 samples, is uncertain. Error bars are +/− 2 standard errors about the mean. After Staley 1992.

Morphological change or stasis in *Melanopsis* sometimes appears related to environmental and/or ecological events. For instance, coincident with the end of stasis and the beginning of gradual change in the *M. impressa–M. fossilis* lineage is the extinction of the last of the reduced marine paratethyan "holdout" taxa, and a reduction in salinity as recorded by shifts in carbon and oxygen isotopes (Geary et al. 1989). This timing suggests that external physical and/or biotic factors constrained *M. impressa* throughout its long period of stasis, and that the early Pan-

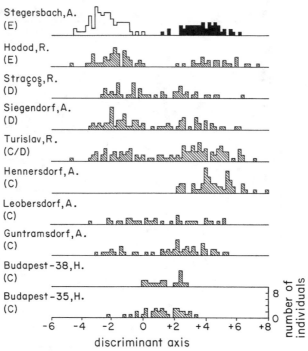

Figure 3.4 Histograms for individuals of *Melanopsis fossilis* and *M. vindobonensis* on the discriminant axis that best separates these two species. The discriminant analysis was carried out on all individuals from Stegersbach, Austria, where the two species are visually distinct (*M. fossilis* shown in white, *M. vindobonensis* in black). This axis was then used to classify individuals from all other samples (stippled). Names at left are sample localities in Austria (A), Romania (R), and Hungary (H). Letters in parentheses below locality names are zones within the Pannonian Stage (see figure 3.1). Size characters were not used in the analysis because individuals of all ontogenetic ages were included. Characters describing shouldering, shell shape, and relative thickness of the apertural lip correlated most strongly with the discriminant function. After Geary 1992.

nonian opening of new habitat facilitated changes in this lineage. In another example, the end of incomplete reproductive isolation between *M. fossilis* and *M. vindobonensis* is coincident with the extinction of several other melanopsid species, suggesting that environmental deterioration may have disrupted this long-standing situation (see Geary 1992). Uncovering the adaptive basis for particular morphological changes is difficult, however; iterative gradual changes in shouldering that occur in the Pannonian and again in the Pontian Stage do not relate to any known physical or chemical factors (Geary 1990a; Staley 1992; Staley et al. submitted; see also Allmon and Geary 1986).

Marginellid Gastropods

Working with the marine gastropod *Prunum* (Family Marginellidae), Ross Nehm and I documented a gradual morphological transition during a relatively rapid speciation event (Nehm and Geary 1994). The ancestral species, *Prunum coniforme*, has a geologic range of more than 11 million years in the Mio-Pliocene deposits of the Dominican Republic and Jamaica. The transition between *P. coniforme* and its descendant *P. christineladdae* is preserved in the well-studied Rio Gurabo section of the northern Dominican Republic, the region to which *P. christineladdae* was endemic.

Typical members of the two species separate well visually and with discriminant analysis; *P. coniforme* is consistently more robust in shell shape, possesses a lower spire, and has denticulations on the inner apertural lip that are lacking in *P. christineladdae* (figure 3.5). Through a 22 meter interval of the Rio Gurabo section, designated here as zones E and F, gradual change from *P. coniforme* to *P. christineladdae* occurs via a sequence of morphological intermediates. The series of histograms in figure 3.5 shows this temporal pattern along the discriminant axis that best separates ancestral and descendant forms. Based on estimates of sedimentation rate, the transitional interval represents approximately 73,000–275,000 years, or 0.6–2.5% of the entire duration of the ancestral species (Nehm and Geary 1994). Thus, the transitional interval, though not a geologic instant, is nevertheless quite rapid in comparison with the entire range of *P. coniforme*.

The appearance of *P. christineladdae* in the Rio Gurabo section marks its first appearance anywhere; the species does not occur below level E in the Rio Gurabo, nor is it found in other sections in horizons predating this interval, including some from deeper water environments (see following).

Prunum species in the Dominican sections appear to track particular depth ranges (figure 3.6). The Rio Gurabo section, represented by the broad line on the right of figure 3.6, is a deepening upward sequence. Significantly, the species-level transition that we observe in the Rio Gurabo section occurs during an interval of accelerated deepening, as indicated by a host of changes in the accompanying faunas.

These examples, together with numerous others documented by many workers, demonstrate that a wide variety of evolutionary tempos occurs

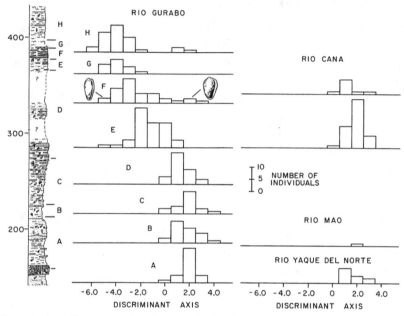

Figure 3.5   Histograms for all individuals of *Prunum coniforme* and *P. christineladdae* on the discriminant axis that best separates these two species. Samples are arranged stratigraphically; A–H represent stratigraphic intervals from the Rio Gurabo section shown at left (numbers mark meters of section). Lower and upper Rio Cana samples are roughly equivalent to E and F, respectively, on the Gurabo section. Samples from the Rio Yaque del Norte and Rio Mao are stratigraphically lower than A. Cartoons at level F show typical morphologies of *P. christineladdae* (left) and *P. coniforme* (right). Levels A–D contain only *P. coniforme*; level E contains mostly individuals of intermediate morphology; level F contains typical individuals of both species plus intermediates; levels G and H contain predominantly *P. christineladdae*. Specimens from the Rio Cana, Rio Mao, and Rio Yaque del Norte are all typical *P. coniforme*. Eight variables describing shell shape, apertural shape, and apertural lip characteristics were used in the analysis. After Nehm and Geary 1994.

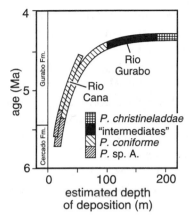

Figure 3.6 Generalized cartoon of estimated water depth for the Rio Cana and Rio Gurabo deposits with time, showing *Prunum* species ranges. Formations have different thicknesses in the two river valleys; the Cercado Formation is approximately 280 m thick in the Rio Cana section and 150 m thick in the Rio Gurabo section. The Gurabo Formation is approximately 300 m thick in the Rio Cana section and 420 m thick in the Rio Gurabo section. After Nehm and Geary 1994.

in the fossil record. Although the level of resolution and stratigraphic completeness vary widely among studies, the distribution of tempos does not appear to be bimodal (geologically instantaneous versus gradual).

Evolutionary Tempo as an Expression of Organism-Environment Interactions

What we measure as evolutionary tempo is in fact the morphological outcome of a complex set of developmental interactions and interactions between organisms and their environment. Of course, "organisms" and the "environment" do not constitute two distinct entities, with environmental changes of a certain magnitude resulting in organismal changes of some appropriately scaled magnitude. Instead, as discussed by Levins and Lewontin (1985) and Kitchell (1990; see also Lewontin 1982; Fisher 1985; Geary 1990a), organisms (or their evolutionary lineages) and the abiotic and biotic environment each represent complex systems whose response to the other is characterized by numerous feedback loops, internal trade-offs, thresholds, and a variety of other contingencies.

Organism-environment interactions are assumed to be the principal driving force for evolutionary change, but the nature of these interactions is not well understood. In an effort to clarify the dynamics of organism-environment interaction, Kitchell and co-workers modeled predator-prey interactions as coupled, nonlinear systems (DeAngelis, Kitchell, and Post 1985; Kitchell 1990). They chose to model predatory naticid gastropods and their bivalve prey because this system can be studied empirically on both ecological and evolutionary time scales. In particular, they modeled the energy allocation strategies of bivalve prey (growth versus reproduction; growth in shell thickness versus length) in response to different levels of predation. Their simulations resulted in a variety of modes of change, from smooth and continuous to abrupt or even static, despite the fact that neither the direction of selection nor the rules of the interaction changed (see Plotnick and McKinney [1993] for discussion of a comparable situation at the ecosystem level).

The tempo of morphologic evolution will be subject to the same kinds of internal complexities and contingencies as the predator prey systems modeled by DeAngelis, Kitchell, and others. For example, gradual change in an environmental parameter may be perceived by a lineage as abrupt, as some threshold value is crossed. Or, relatively major environmental changes sometimes appear to have no influence whatsoever on organisms (Williams 1992). Prothero has investigated the response of a wide variety of mammal lineages to the dramatic climate changes that occurred during the late Eocene–early Oligocene (see Prothero and Berggren 1992). His results show that some lineages go extinct during this time, but the majority appear unaffected by the environmental changes and persist with no appreciable (or directional) morphologic response. (For a variety of other responses or nonresponses to environmental change see Malmgren and Berggren 1987, Sorhannus et al. 1988, and McGhee 1991 and 1992). Furthermore, a host of biological factors may modify the tempo of response to directional pressure, including the number of loci involved, epistatic interactions, and the amount of variation present (Charlesworth 1984a; Levinton 1988; Williams 1992).

When evolutionary tempo is considered as the outcome of complex organism-environment interactions, one expects a wide spectrum of temporal patterns. One may ask whether this spectrum would in fact extend across geologic, as well as ecologic, time scales; are there absolute limits to the duration of a transition, a threshold below which changes must take place or the impetus for change can not be maintained? The cases described above (as well as others, such as Malmgren et al. 1983; Chaline and Laurin 1986; Lazarus 1986; Sheldon 1987) argue against an absolute temporal limit for the duration of a transition. In general, gradual tempos appear to be common enough to merit attention in our explanations of the process of change.

The Importance of Gradual Change

When Eldredge and Gould (1972) initially formulated their theory, they argued that a punctuated pattern was the logical geologic outcome of standard models of allopatric speciation (see also Gould 1982). Nothing in speciation theory during the last twenty years challenges this basic notion; evolutionary changes can happen quite rapidly, and most of the time their geologic expression will be abrupt. As Gould and Eldredge (1986) have pointed out, a great many biologists now insist that they knew this all along. Also fundamental to punctuated equilibria is the stability of species in geologic time; the prevalence of stasis has demonstrated that species often remain stable for millions of years.

The point to be emphasized here is that situations other than stasis can also be maintained for long time periods (see also Bakker 1983). For example, apparent incomplete reproductive isolation between *M. fossilis* and *M. vindobonensis* persisted for more than one million years. Various other cases in which hybridization appears to have been ancient or long-lasting have been described (Mayr 1942; White et al. 1967; Short 1972; Hall and Selander 1973; Hunt and Selander 1973; Woodruff 1979; Woodruff and Gould 1980; Gould and Woodruff 1986). Furthermore, cases of gradualism indicate that conditions consistent enough to result in net directional morphological change may persist for millions of years.

Change over long intervals is not necessarily change at very slow generation-by-generation rates, because the calculation of rate is so dependent on the time interval over which the rate is measured (Gingerich 1983). If we could break down these long intervals of gradual change into shorter, well-resolved segments of time, we would undoubtedly see many fluctuations in both the rate and the direction of evolutionary change (Lande 1976; Raup and Crick 1981; Bell and Haglund 1982; Charlesworth 1984b; Bell et al. 1985; Stanley and Yang 1987).

This standard explanation for change that occurs over long intervals obscures a variety of forces and constraints that would be interesting to untangle. Cases of geologically gradual change ought to command more attention because they indicate that, even if not continuous, some process has prevailed for a considerable time span. Gradual change is compelling for reasons similar to those that justify the statement "stasis is data"— something is going on during that interval that may tell us more about the nature of species. In particular, even if intermittent, change is nonetheless sustained over these long intervals. The mechanisms that sustain the change, as well as those that inhibit it, are of interest.

As with stasis, it seems likely that changes spanning long time periods often reflect constraint. That is, a part of what is causing the apparently slow pace of change may reflect basic limitations in the response of the lineage or some sort of evolutionary lag. One of the messages of punctuated equilibria is that species are not especially malleable. The same message might also be derived from geologically gradual change; if species are morphologically flexible, why should change take hundreds of thousands of years or more? Cases of gradual change may indicate that the reaction of a lineage to some stimulus is not necessarily all-or-none (i.e., speciation versus extinction or stasis).

Alternatively, some cases of gradual change may accurately reflect the pace of environmental change. Directional selection pressures may be intermittent for long time intervals. It is unrealistic to assume that selection pressures similar to those in laboratory experiments would continue unabated in the wild for thousands of generations (Levinton 1988; Williams 1992).

For these reasons, what is happening during extended intervals of "slow" change might be fertile ground for more detailed investigations. Sorting out internal constraints from intermittent selection pressures is a big order, but the fossil record contains significant clues. Most important among these, and underutilized to date, is detailed information on geographic variation and on the sedimentary environment. I will focus here on the former.

Geographic Variation

The most obvious means by which paleontologists can investigate intrinsic versus extrinsic effects on a species or the dynamic nature of species over time is through study of geographic variation. Only by understanding the morphologic limits of a species across its geographic range can we hope to interpret morphologic change over time (see Bell, Baumgartner, and Olson 1985; Wei and Kennett 1988). The following examples illustrate this point.

Cardiid Bivalves

A study of tempo and mode in the cardiid bivalve *Pleuriocardia* from the Cretaceous of the Western Interior of North America (Geary 1987) was not particularly conclusive with respect to identifying clear examples of either punctuated or gradual change, but it illustrates the importance of geographic information for interpreting temporal patterns. The sequences in figure 3.7 represent geographic and temporal variation within an ancestral-descendant species pair. The better-sampled descendant species, *P. pauperculum* (indicated by black circles), exhibits minor fluctuations in a number of characters, some of which suggest directional change, but the "trends" are difficult to distinguish from a random walk (Bookstein, Gingerich, and Kluge 1978; Raup and Crick 1981; Charlesworth 1984a). For instance, the total number of axial ribs per specimen is plotted in figure 3.7a. A minor shift is apparent from the bottom to the top of the section, but the overall magnitude of the temporal change is small relative to geographic variation among roughly contemporaneous samples at approximately 89 Ma and again at approximately 88 Ma. The same pattern is illustrated in figures 3.7b and 3.7c for two characters that measure shell shape.

For these characters then, the magnitude of geographic variation among roughly contemporaneous samples is approximately comparable to the apparent overall temporal shift within *P. pauperculum*. The relevance of this to interpreting temporal sequences in general is evident; samples from a single locality can not adequately represent species variability.

Muricid Gastropods

Gary Gianniny and I examined geographic and temporal variation in the neogastropod *Acanthina spirata* from the California and Baja California coast (Gianniny and Geary 1992). The only mode of dispersal for species of *Acanthina* is through active movement of individuals; fertilization is

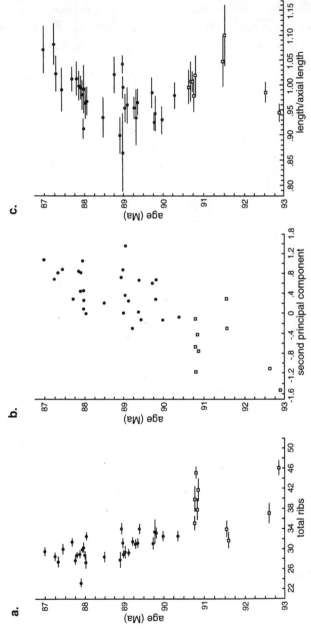

Figure 3.7 Plots of selected characters over time in the Cretaceous bivalve *Pleuriocardia subcurtum* (open squares) and *P. pauperculum* (black circles) in the Western Interior of North America. Error bars represent +/− two standard errors about the mean. (a) Total number of axial ribs; (b) Sample mean scores on the second axis of a principal components analysis; (c) Ratio of shell length to axial length. After Geary 1987.

internal, and juveniles emerge directly from egg cases attached to rocky surfaces. Modern populations of *A. spirata* range from northern California to central Baja California; we measured individuals from 12 modern samples and 22 fossil samples. Both modern and fossil samples ranged from near San Francisco Bay (approximately 37.7°N) southward to Punta Baja (approximately 30°N, roughly one-third of the way down the length of the Baja California peninsula). We found that modern samples were consistently different from Late Pleistocene samples with respect to the development of shell shouldering. Modern samples from Los Angeles provide an exception to this pattern, however, in that they appear more similar to the fossils (figure 3.8).

During the Pleistocene, the latitudinal range of *A. spirata* moved up and down the coast in response to climate fluctuations. It seems likely that the morphological patterns we see today reflect these range expansions and contractions. Specifically, the similarity of modern Los Angeles

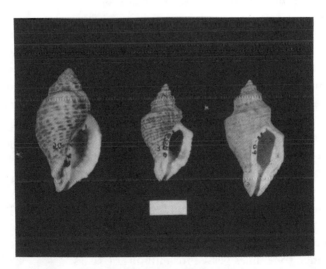

Figure 3.8 Typical specimens of the neogastropod *Acanthina spirata*, sampled along the California and Baja California coast. Note the characteristic differences in shouldering between fossil (right) and modern non-Los Angeles (left) specimens. The modern Los Angeles specimen (center) more closely resembles typical fossil shouldering morphology. Scale bar is 1 cm. Source: Gianniny and Geary 1992.

samples to those everywhere in the Late Pleistocene suggests that populations have remained in the Los Angeles area more continuously than elsewhere. Modern samples may have spread north (and south) from Los Angeles, and in so doing, established their characteristic morphology.

Summary of Geographic Variation

These examples illustrate how knowledge of geographic variation is essential to knowing the morphological limits within a species, and to interpreting the history of a species. Such issues are central to understanding what species are and how they change. From the beginning, Eldredge and Gould have emphasized the importance of geographic variation, but many studies since have lacked geographic information. As we move away from simply cataloging cases of punctuated versus gradual change to the more difficult issues of what determines stasis or species-level change, we cannot ignore geographic variation.

Organisms (and their evolutionary lineages) and the abiotic and biotic environment are complex systems whose responses to one another are characterized by feedback loops, trade-offs, thresholds, and a host of other contingencies. Because evolutionary tempo is the outcome of organism-environment interactions, one expects a wide spectrum of temporal patterns. This spectrum ranges across geologic as well as ecologic time scales.

Stasis is not the only tempo that can be maintained over long time periods. In particular, cases of geologically gradual change are compelling by virtue of their apparent slowness. They indicate that a species response to a stimulus may be incremental or sporadic, rather than simply all or none. Furthermore, cases of geologically gradual change indicate that, even if intermittent, conditions consistent enough to result in net directional change sometimes persist for millions of years.

Information on geographic variation is critical to understanding the limits of a species and interpreting temporal patterns. In addition, we can get a better sense of the environmental context for evolutionary change by using more of the tools of sedimentary geology.

Studies of tempo and mode in the fossil record remain central to the paleontological study of evolution. Over the last twenty years, the specific questions being addressed have shifted; newer concerns will prove more challenging. Using fossil data we can investigate the nature of species, including the nature of constraint, and the interaction of organism and environment. Continuing in the tradition established by Eldredge and Gould, paleontologists ought to have at least as much to say about these topics as do population or developmental biologists.

Acknowledgments I thank Warren Allmon, Robert Bleiweiss, Charles Byers, and Douglas Erwin for useful comments on the manuscript.

References

Allmon, W. D. and G. H. Geary. 1986. A pattern of homeomorphy in diverse lineages of gastropods. *Fourth North American Paleontological Convention Abstracts.* A1.

Bakker, R. T. 1983. The deer flees, the wolf pursues: Incongruencies in predator-prey coevolution. In D. J. Futuyma and M. Slatkin, eds., *Coevolution,* pp. 350–382. Sunderland, Mass.: Sinauer.

Bell, M. A., J. V. Baumgartner, and E. C. Olson. 1985. Patterns of temporal change in single morphological characters of a Miocene stickleback fish. *Paleobiology* 11:258–271.

Bell, M. A. and T. R. Haglund. 1982. Fine-scale temporal variation of the Miocene stickleback *Gasterosteus doryssus. Paleobiology* 8:282–292.

Bookstein, F., P. D. Gingerich, and A. Kluge. 1978. Hierarchical linear modeling of the tempo and mode in evolution. *Paleobiology* 4:120–134.

Chaline, J. and B. Laurin. 1986. Phyletic gradualism in a European Plio-Pleistocene *Mimomys* lineage (Arvicolidae, Rodentia). *Paleobiology* 12:203–216.

Charlesworth, B. 1984a. The cost of phenotypic evolution. *Paleobiology* 10:319–327.

———. 1984b. Some quantitative methods for studying evolutionary patterns in single characters. *Paleobiology* 10:308–318.

Cheetham, A. H. 1986. Tempo of evolution in a Neogene bryozoan: Rates of morphologic change within and across species boundaries. *Paleobiology* 12:190–202.

DeAngelis, D. L., J. A. Kitchell, and W. M. Post. 1985. The influence of naticid predation on evolutionary strategies of bivalve prey: Conclusions from a model. *American Naturalist* 126:817–842.

Eldredge, N. and S. J. Gould. 1972. Punctuated equilibria: An alternative to phyletic gradualism. In T. J. M. Schopf, ed., *Models in Paleobiology,* pp. 82–115. San Francisco: Freeman Cooper.

Fisher, D. C. 1985. Evolutionary morphology: Beyond the analogous, the anecdotal, and the ad hoc. *Paleobiology* 11:120–138.

Fortey, R. A. 1985. Gradualism and punctuated equilibria as competing and complementary theories. *Special Papers in Palaeontology* 33:17–28.

Geary, D. H. 1987. Evolutionary tempo and mode in a sequence of the Upper Cretaceous bivalve *Pleuriocardia. Paleobiology* 13:140–151.

———. 1990a. Evaluating intrinsic and extrinsic factors in the evolution of *Melanopsis* in the Pannonian Basin. In R. M. Ross and W. D. Allmon, eds., *Causes of Evolution, a Paleontological Perspective*, pp. 305–321. Chicago: University of Chicago Press.

———. 1990b. Patterns of evolutionary tempo and mode in the radiation of *Melanopsis* (Gastropoda; Melanopsidae). *Paleobiology* 16:492–511.

———. 1992. An unusual pattern of divergence between two fossil gastropods: Ecophenotypy, dimorphism, or hybridization? *Paleobiology* 18:93–109.

Geary, D. H., J. A. Rich, J. W. Valley, K. Baker. 1989. Isotopic evidence for salinity changes in the Late Miocene Pannonian Basin: Effects on the evolutionary radiation of melanopsid gastropods. *Geology* 17:981–985.

Gianniny, G. L. and D. H. Geary. 1992. Geographic and temporal variation in shell morphology of *Acanthina* species from California and northern Baja California. *The Veliger* 35:195–204.

Gingerich, P. D. 1983. Rates of evolution: Effects of time and temporal scaling. *Science* 222:159–161.

Gould, S. J. 1982. The meaning of punctuated equilibrium and its role in validating a hierarchical approach to macroevolution. In R. Milkman, ed., *Perspectives on Evolution*, pp. 83–104. Sunderland, Mass.: Sinauer.

Gould, S. J. and N. Eldredge. 1986. Punctuated equilibrium at the third stage. *Systematic Zoology* 35:143–148.

Gould, S. J. and D. S. Woodruff. 1986. Evolution and systematics of *Cerion* (Mollusca: Pulmonata) on New Providence Island: A radical revision. *Bulletin of the American Museum of Natural History* 182:389–490.

Hall, W. P. and R. K. Selander. 1973. Hybridization of karyotypically differentiated populations in the *Sceleporus grammicus* complex (Iguanidae). *Evolution* 27:226–242.

Hunt, W. G. and R. K. Selander. 1973. Biochemical genetics of hybridisation in European house mice. *Heredity* 31:11–33.

Kitchell, J. A. 1990. The reciprocal interaction of organism and effective environment: Learning more about "and." In R. M. Ross and W. D. Allmon, eds., *Causes of Evolution, a Paleontological Perspective*, pp. 151–169. Chicago: University of Chicago Press.

Lande, R. 1976. Natural selection and random genetic drift in phenotypic evolution. *Evolution* 30:314–334.

Lazarus, D. 1986. Tempo and mode of morphologic evolution near the origin of the radiolarian lineage *Pterocanium prismatium. Paleobiology* 12:175–189.

Levins, R. and R. C. Lewontin. 1985. *The Dialectical Biologist*. Cambridge: Harvard University Press.

Levinton, J. 1988. *Genetics, Paleontology, and Macroevolution*. Cambridge: Cambridge University Press.

Lewontin, R. C. 1982. Prospectives, perspectives and retrospectives. *Paleobiology* 8:309–313.

Malmgren, B. A. and W. A. Berggren. 1987. Evolutionary changes in some Late Neogene planktonic foraminiferal lineages and their relationships to paleoceanographic changes. *Paleoceanography* 2:445–456.

Malmgren, B. A., W. A. Berggren, and G. P. Lohmann. 1983. Evidence for punctuated gradualism in the Late Neogene *Globorotalia tumida* lineage of planktonic foraminifera. *Paleobiology* 9:377–389.

Mayr, E. 1942. *Systematics and the Origin of Species.* New York: Columbia University Press.

McGhee, G. R. 1991. Extinction and diversification in the Devonian Brachiopoda of New York State: No correlation with sea level? *Historical Biology* 5:215–227.

———. 1992. Evolutionary biology of the Devonian Brachiopoda of New York State: No correlation with rate of change of sea level? *Lethaia* 25:165–172.

Müller, P. 1992. Personal communication.

Nehm, R. H. and D. H. Geary. 1994. A gradual morphological transition during a rapid speciation event in marginellid gastropods (Neogene; Dominican Republic). *Journal of Paleontology* 68:787–795.

Plotnick, R. E. and M. L. McKinney. 1993. Ecosystem organization and extinction dynamics. *Palaios* 8:202–212.

Prothero, D. R. and W. A. Berggren, eds. 1992. *Eocene-Oligocene Climatic and Biotic Evolution.* Princeton: Princeton University Press.

Raup, D. M. and R. E. Crick. 1981. Evolution of single characters in the Jurassic ammonite *Kosmoceras. Paleobiology* 7:200–215.

Sheldon, P. R. 1987. Parallel gradualistic evolution of Ordovician trilobites. *Nature* 330:561–563.

Short, L. L. 1972. Hybridization, taxonomy and avian evolution. *Annals of the Missouri Botanical Garden* 59:447–453.

Sorhannus, U. 1990. Tempo and mode of morphological evolution in two Neogene diatom lineages. *Evolutionary Biology* 24:329–370.

Sorhannus, U., E. J. Fenster, L. H. Burckle, and A. Hoffman. 1988. Cladogenetic and anagenetic changes in the morphology of *Rhizosolenia praebergonii* Mukhina. *Historical Biology* 1:185–205.

Staley, A. W. 1992. Patterns of morphologic change and iterative evolution in the gastropod genus *Melanopsis* from the Late Miocene Pannonian Basin. M.S. thesis, University of Wisconsin, Madison.

Staley, A. W., D. H. Geary, P. Müller, and I. Magyar. Submitted. An iterative evolutionary pattern in the gastropod genus *Melanopsis.*

Stanley, S. M. and X. Yang. 1987. Approximate evolutionary stasis for bivalve morphology over millions of years: A multivariate, multilineage study. *Paleobiology* 13:113–139.

Wei, K. Y. and J. P. Kennett. 1988. Phyletic gradualism and punctuated equilibrium in the late Neogene planktonic foraminiferal clade *Globoconella. Paleobiology* 14:345–363.

White, M. J. D., R. E. Blackith, R. M. Blackith, and J. Cheney. 1967. Cytogenetics of the *viatica* group of morabine grasshoppers. I. The "coastal" species. *Australian Journal of Zoology* 15:263–302.

Williams, G. C. 1992. *Natural Selection: Domains, Levels, and Challenges.* New York: Oxford University Press.

Woodruff, D. S. 1979. Postmating reproductive isolation in *Pseudophryne* and the evolutionary significance of hybrid zones. *Science* 203:561–563.

Woodruff, D. S. and S. J. Gould. 1980. Geographic differentiation and speciation in *Cerion:* A preliminary discussion of patterns and processes. *Biological Journal of the Linnean Society* 14:389–416.

4

Phylogenetic Patterns as Tests of Speciation Models

Peter J. Wagner and Douglas H. Erwin

Most paleobiological discussions about phylogeny have focused on particular transitions from an ancestral species to a descendant one, or on a few transitions within a single taxon (e.g., Gingerich 1985). Although valuable, these studies represent individual examples of a pattern of speciation rather than tests of the generality of any particular mode. Moreover, adequately documented speciation events are very rare. Consequently, even if the investigators have a hypothesis about a clade's phylogeny, they usually lack any direct evidence of the evolutionary processes that produced that phylogeny.

A different approach to testing hypotheses about patterns of speciation among fossil species involves a strategy of examining phylogenetic patterns. Different speciation models predict different relationships between speciation and the survival of ancestral morphologies. More specific models also postulate different associations between species-level traits (e.g., temporal and geographic ranges) and the likelihoods of leaving descendants. Therefore, several additional predictions stem from the hypothesis that a particular mode of speciation (vicariance, peripheral isolation, etc.) predominates within a clade. These include which patterns of speciation (anagenesis, bifurcation, cladogenesis) and phylogenetic branching (polytomies, pectinate branches, etc.) should be most common, and the relative likelihoods of sampling plesiomorphic and apomorphic species from the fossil record.

Because hypotheses about modes of speciation ultimately make predictions about the phylogenetic geometries of clades, phylogenetic patterns can be used to corroborate or refute hypotheses predicting that particular modes of speciation were dominant among closely related taxa. Here we will provide examples of the phylogenetic patterns produced by cladogenesis, anagenesis, and bifurcation. Although reticulation certainly is common among plants and possibly more common among animals than generally realized (e.g., Smith 1992), it is much more difficult to recognize in the fossil record. We therefore will not discuss examples of reticulation.

We will use two previously published computer simulations of phylogeny to highlight how different assumptions about speciation patterns produce very different phylogenetic patterns. We then use some basic assumptions derived from the simulated results to examine two examples from the fossil record. We also discuss why cladograms alone are insufficient to describe phylogenetic patterns without hypothesizing about ancestor-descendant relationships and evaluating temporal or geographic data. Finally, we examine why phylogenies must be interpreted without preconceptions about how speciation proceeds and the implications of this for phylogenetic systematics and comparative biology.

Definitions of Terms

Because patterns of speciation have been defined differently by various authors, we will define exactly how we are using each term. The original definition of "cladogenesis" (Rensch 1959, p. 97) included two patterns: (1) ancestral lineages giving rise to two daughter lineages, with the ancestral morphology disappearing, and (2) ancestral lineages "branching" into new daughter lineages with no necessary change co-occurring in the ancestral morphology (see figure 4.1). Some modern definitions (e.g., Maddison and Slatkin 1991; Nixon and Wheeler 1992) restrict cladogenesis to the first pattern. Other definitions (e.g., Raup 1977; Raup and Gould 1974) restrict cladogenesis to the second.

The definition that restricts cladogenesis to the first pattern incorporates elements of anagenesis (i.e., the replacement of an ancestral morphology by a derived morphology within a single lineage) because it assumes that the ancestral morphology becomes "pseudoextinct." Combining anagenesis and cladogenesis is confusing; therefore we restrict our own definition of cladogenesis to the second pattern and use "bifurcation" to describe the first pattern. In addition, although some workers do not consider speciation to occur via anagenetic evolution ("phyletic evolution" versus "speciation," *sensu* Larson 1989), we will consider bifurcation a pattern of speciation here.

Table 4.1 lists some modes of speciation and the patterns of speciation and phylogenetic topologies that they predict. As with patterns of speciation, the definitions of modes of speciation have varied. Our "vicari-

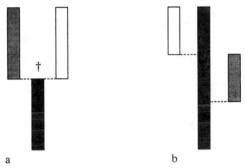

a   b

Figure 4.1   The two patterns encompassed by Rensch's (1959) definition of "cladogenesis." (a) Ancestral species become pseudo-extinct (†) during a bifurcation into two derived daughter lineages. (b) Derived species branch off from ancestral ones, with the ancestral species conserving a nonderived morphology. Here we use *cladogenesis* to refer to the second pattern only, while we define the first pattern as *bifurcation*.

Table 4.1
Modes and Patterns of Speciation

Mode	Speciation Pattern (typical)	Phylogenetic Pattern
Selection-Driven Divergence	Bifurcation	Symmetric
Sympatry	Bifurcation	Symmetric
Parapatry	Bifurcation	Symmetric
Vicariance	Bifurcation	Symmetric
Anagenesis	Pseudo-extinction	Pectinate
Peripheral Isolation	Cladogenesis	Polytomy
Hybridization	Reticulation	?

SOURCES: "symmetric" = "balanced" of Heard 1992; "pectinate" = "Hennigian comb" of Panchen 1992 and "unbalanced" of Heard 1992.
 NOTES: Hypothesized modes of speciation are matched with the typical speciation patterns each predicts and with the phylogenetic pattern expected if that mode predominates within a clade.

ance" encompasses both the "type II" and "type III" allopatry of Brooks and McLennan (1991). We use "peripheral isolation" (*sensu* Mayr 1963 = "peripatry" of Mayr 1982) for Brooks and McLennan's "type I" allopatry. Because definitions of allopatry have encompassed very different patterns of speciation, we do not use that term here.

Effects of a Priori Assumptions of Interpreting Phylogenetic Pattern

If we are to use phylogenies to test hypotheses about speciation patterns, then the methods of phylogenetic reconstruction must not assume a particular speciation model. Many workers appear to think that cladistic (parsimony) analyses require a bifurcating pattern of speciation (e.g., Lorenzen and Sieg 1991). This is not entirely correct. For example, the matrix shown in figure 4.2 yields a single most parsimonious tree with a polytomy. Nevertheless, many phylogenetic systematists do assume that bifurcation is the standard pattern of speciation (e.g., Hennig 1966; Maddison 1989; Slowinski and Guyer 1989a,b; Nixon and Wheeler 1992; Heard 1992) when interpreting cladograms. Such systematists would not interpret the polytomy in figure 4.2 as one species giving rise to four but as an artifact of ignorance. The assumption (often explicit) is that most, if not all, polytomies are the result of inadequate data, and that inclusion of more characters would "resolve" some bifurcating pattern.

The assumption of bifurcation obviously imposes a particular pattern of speciation on the interpretation of evolutionary patterns and ignores speciation models that allow a single species to produce multiple daughter

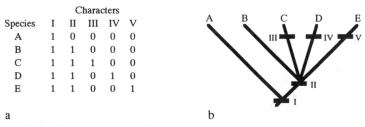

a   b

Figure 4.2 A data matrix (a) whose most parsimonious solution produces a polytomy, shown in (b). Without other data, it cannot be determined whether this pattern is due to a single species giving rise to several others or to insufficient data. However, this topology should be found by any parsimony algorithm that does not assume bifurcation.

species. To avoid such assumptions, we will take a cue from recent methodological works in comparative biology (e.g., Harvey and Pagel 1991) and assume that the cladograms presented here are more or less correct even if they illustrate polytomies. We therefore accept *some* polytomies as signal rather than noise.

Models of Speciation and the Phylogenetic Patterns They Predict

Figure 4.3 shows the phylogeny of a bifurcating lineage. Commonly cited mechanisms for bifurcation include selection-driven divergence (e.g., Darwin 1859; Gingerich 1976), sympatry (e.g., Maynard Smith 1966; Wake, Yanev, and Frelow 1989), parapatry (e.g., Lande 1982), and vicariance (e.g., Lynch 1989; Brooks and McLennan 1991). Although it has been noted that a descendant lineage could retain a plesiomorphic morphology (Nixon and Wheeler 1992), this is not the prediction of selection-driven divergence nor is it a likely prediction of vicariance (e.g., see Brooks and McLennan 1991).

Slowinski and Guyer (1989a,b; see also Guyer and Slowinski 1991) used three algorithms to generate cladograms, one of which generated phylogenies using only bifurcation. These simulations assumed that all species were equally likely to leave descendants, so extrinsic features such as temporal and geographic ranges were not relevant to the outcomes. Also, the possibility of true extinction was ignored. If all species are assumed to be extant and only bifurcation occurred, only three evolutionary trees are possible (figure 4.4a–c). Each of these possible trees can be depicted as a cladogram (d from a, e from b, f from c). These cladograms each assume that the ancestral species became "pseudoextinct" after each speciation event. If, however, these assumptions are relaxed and fossil taxa and polychotomies are allowed, then the addition of ancestral species produces very different cladistic patterns for the same original trees (cladograms g–i).

If one predicts (or merely assumes) that bifurcating patterns of speciation predominated within a clade, then one also is hypothesizing that the apparent extinction of plesiomorphic species should be statistically indistinguishable from the apparent originations of daughter lineages. (For discussions on statistical tests of the apparent origins and extinctions of fossil species, see Marshall 1990 and this volume.) Bifurcation models also predict that two derived sister species should share statistically in-

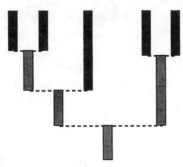

Figure 4.3 A bifurcating phylogenetic pattern. Although this pattern sometimes is termed "cladogenesis," the pattern shown here is equally anagenetic, as
ancestral morphologies are replaced by descendant morphologies. Therefore,
we have reserved the term *cladogenesis* for a different phylogenetic pattern. In
this and in all following figures, ancestral species are given in gray whereas
those without descendants are given in black.

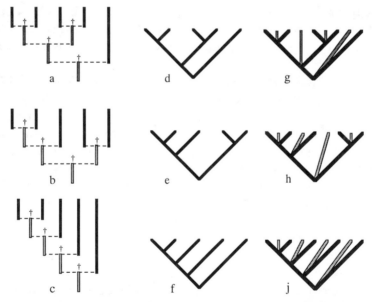

Figure 4.4 Three possible phylogenies (a–c) for five species if bifurcation is
the only mode of speciation. Note that we assume no extinction and complete
sampling of the taxon. "†" denotes the pseudo-extinction of the ancestral lineage at each bifurcation. (d–f) Three cladograms for the *final five species* produced in Slowinski's and Guyer's (1989) simulations. These simulate sampling
species from one time line (e.g., the present), given the assumption that ancestral morphologies cannot be sampled. (g–j) The previous three cladograms with
ancestral species included.

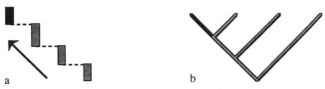

Figure 4.5 Phylogeny of anagenetic change (a) with corresponding cladogram (b). Although each "species" is represented by a vertical band in (a), this is meant solely to represent the temporal range of a morphotype, not to imply morphologic stasis.

distinguishable first appearances in the fossil record. Although some workers have considered the latter statement to be true of all sister taxa (e.g., Cracraft 1981), it is necessarily true only for cases in which speciation actually followed a bifurcating pattern.

Figure 4.5a depicts a lineage in which only anagenesis occurs. A cladogram of "morphospecies" from this lineage should be completely pectinate (*sensu* Slowinski and Guyer 1989a,b; — "unbalanced" of Heard 1992) and all of the lineages save for one would be plesiomorphic (figure 4.5b). Both gradual and punctuated models can predict anagenetic speciation (Wright 1931, 1932; see Jablonski 1986, Provine 1989, and MacLeod 1991). Boucot (1978) has argued that this is indeed the dominant pattern observed in the fossil record. Predictions that anagenetic modes of speciation predominated within a clade thus posit that plesiomorphic species should be fairly common and that the temporal ranges of those species should cease where the ranges of (relatively) apomorphic species begin. One might also predict general congruence between the geographic ranges of plesiomorphic and apomorphic species.

Speciation models that predict cladogenesis include models of peripheral isolation (e.g., Mayr 1963; Eldredge 1971; Eldredge and Gould 1972) and some interpretations of the shifting balance theory (e.g., Wright 1982; Eldredge, this volume). These models do not predict that the evolution of a daughter species necessarily coincides with the extinction of the ancestral species. Furthermore, because the factors affecting speciation and resistance to extinction in some models (i.e., the number of demes and their spatiotemporal distributions) are not necessarily altered by speciation (e.g., Wright 1982; Lande 1980, 1986), these models may even predict that the ancestral lineage is more likely to speciate in the future than is the descendant one.

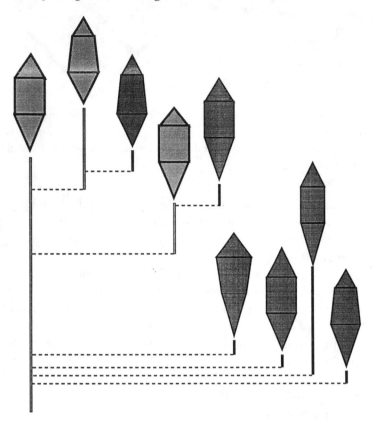

Figure 4.6   A segment of the "triloboid" phylogeny taken from the MBL simulations and stereotyping cladogenetic evolution as defined in this chapter.

As already noted, we use "bifurcation" to distinguish one of the two patterns of speciation included within the term "cladogenesis" that is used (too broadly, in our view) by some systematists. Although Maddison (1989) referred to "multiple speciation," where one species gives rise to several descendants, he discussed it as an event (i.e., trifurcation, quadrifurcation, etc.) and thus a special case of processes that produce bifurcation (e.g., "type III" allopatry of Brooks and McLennan 1991). Nixon and Wheeler (1992) acknowledged that one daughter lineage can be indistinguishable from the ancestral morphology, which leaves a pattern indistinguishable from our definition of cladogenesis. However, they did not discuss the possibility or likelihood of the plesiomorphic lineage producing additional daughter taxa.

The well-known MBL phylogeny simulation (e.g., Raup and Gould 1974; Raup 1977) used cladogenesis (as we narrowly define it) exclusively as the pattern of speciation. Because probabilities of extinction and speciation were held constant, the past history of a lineage (i.e., age or number of daughter lineages) did not affect its immediate prospects. Figure 4.6 shows a segment of this phylogeny, and figure 4.7a gives the corresponding cladogram. Figure 4.7b–e depicts cladograms for other segments of the "triloboid" phylogeny. An abundance of polytomies reflecting the actual relationships among "triloboid" species clearly predominates. In addition, a strong association exists between the temporal range of a species and the number of descendants it leaves.

If modes of speciation conducive to our restricted definition of cladogenesis predominate within a clade, then an association between the extinction of one species and the appearance of another is not necessarily expected. Based on the MBL simulation, we might expect a positive association between plesiomorphic species and either longer temporal

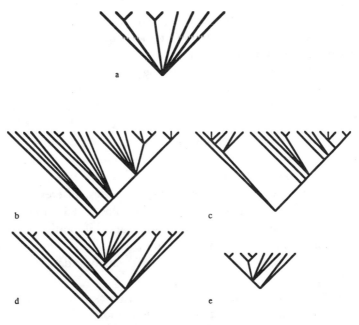

Figure 4.7 Cladogram accompanying figure 4.6; b–e cladograms reflect larger segments of the triloboid phylogeny than does a. Derived from Raup 1977.

ranges *or* whatever factors are conducive to longer temporal durations. Finally, we actually expect polytomies and some "sister taxa" with different times of origin.

Why Cladistic Topology Is Insufficient for Discerning Patterns of Speciation

Although cladograms are often referred to as "phylogenies," a true phylogeny is an evolutionary tree that includes a time element and ancestor-descendant hypotheses (Hennig 1966). Cladograms alone cannot discriminate among patterns of speciation, so discussing speciation requires more than cladistic topologies.

Plotting (or "optimizing") character state transitions onto a cladogram is a fairly common practice. Szalay (1977) suggested that ancestral species should lack apomorphies relative to their immediate descendants and that zero-length branches imply ancestors. Some such criterion is necessary to distinguish when ancestors (and thus information about speciation patterns) may be present. In our own work, if a species is plesiomorphic relative to its sister species (figure 4.8a), then we considered it as a possible ancestor. However, if both sister species are apomorphic (figure 4.8b), then we usually did not consider either species as a possible ancestor. The same reasoning applies to polytomies (figure 4.8c,d). A species plesiomorphic relative to a polytomous node can represent either a "hard" polytomy (Maddison 1989) or a bifurcation that includes the ancestor in the cladogram. "Soft" polytomies, on the other hand, are artifacts of insufficient data and lack any phylogenetic resolution. Among fossil species, soft polytomous nodes cannot include a plesiomorphic species, and thus cannot provide useful data.

Even if a cladogram reveals plesiomorphic species, different speciation patterns still can produce the same cladistic topologies (Paul 1985). Figure 4.9 shows various anagenetic, cladogenetic, and bifurcating topologies that would all produce the same cladogram. The only way to distinguish these patterns of speciation is to incorporate temporal data and convert the cladogram into a phylogenetic tree.

Although temporal data can disprove hypotheses predicting pseudoextinction (i.e., anagenesis or bifurcation), it cannot prove them. The observation that a hypothesized ancestor and its descendant(s) co-occurred

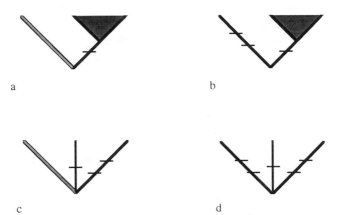

Figure 4.8 (a) Pectinate topology with a plesiomorphic outgroup; (b) pectinate topology with an apomorphic outgroup; (c) polytomy with a plesiomorphic species; (d) polytomy with only apomorphic species. Bars represent apomorphies. Cladograms a and c are potentially informative here· b and d are not.

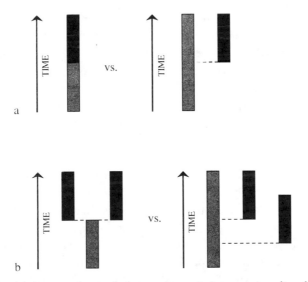

Figure 4.9 (a) Anagenetic vs. cladogenetic speciation patterns that should produce the same cladistic topology; (b) bifurcating vs. cladogenetic speciation patterns that also should produce the same cladistic topology.

is irrefutable evidence that the last appearance of the ancestral species post-dated the first appearance of its descendant(s) (Alroy 1992). In this instance, cladogenesis would be the only possible mode of speciation. However, the observation that a hypothesized ancestor disappears before the first known appearance(s) of its hypothesized descendants is supporting, but not conclusive, evidence of pseudoextinction.

Do We Expect To Sample Ancestral Species from the Fossil Record?

Paleobiologists, including us, have been far more willing to hypothesize the presence of ancestral species than have been other evolutionary biologists. One of the main assumptions behind our arguments is that ancestral species should be fairly common in the fossil record. In this section we discuss whether this assumption is realistic given how factors such as geographic and temporal ranges affect the likelihoods of both leaving descendants and being sampled from the fossil record.

Cladogenetic speciation models as simple as the MBL simulations or as complex as the shifting balance theory predict that species with longer temporal ranges are more likely to leave descendants. In addition, if the basic population dynamics (e.g., larval ecology, dispersal capacity) among species within a clade do not change, species with wider geographic ranges should be more likely to leave descendants. Greater temporal and geographic ranges provide both more demes and more opportunities for those demes to diverge from the rest of the species. Closely related species usually should be subject to similar taphonomic biases. If so, then the main factor affecting the relative probability of any one species being sampled from the fossil record is the amount of sediment in which the species can be preserved (Signor 1985; Valentine 1989). Thus if cladogenesis predominates within a clade, then the expectation is that plesiomorphic species should be more common than apomorphic species in the fossil record.

Modes of speciation predicting bifurcation usually make no predictions about the relative temporal ranges of ancestral and descendant species. However, these models do make predictions about the relative initial geographic ranges of species. This is demonstrated in an example recently published by Brooks and McLennan (1991; see Cracraft 1989 for similar examples). They explained the cladistic relationships among species of

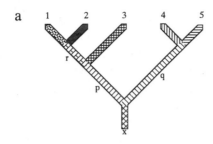

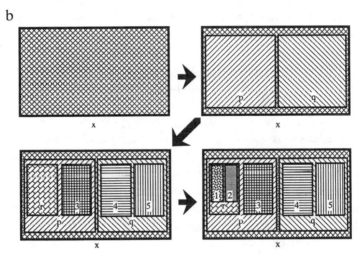

Figure 4.10 Cladogram (a) corresponding to several vicariant events (b) that divided an ancestral geographic range. Adapted from Brooks and McLennan 1991 (figure 4.16, p. 112).

freshwater fish (figure 4.10a) as a series of vicariant events, which are depicted in figure 4.10b. The range of the ultimate ancestor is the widest, with subsequent vicariance events progressively dividing that range. Although Brooks and McLennan noted that ranges can and do expand over time, the null expectation is that the oldest species will have the widest geographic ranges and ancestors thus will have higher likelihoods of being sampled from the fossil record.

Even if we expect to see ancestors in the fossil record, this does not demonstrate that any particular pair of species are ancestor and descendant. It has thus been claimed that the hypothesis that the ancestor has the widest geographic range is ultimately untestable (e.g., Engelmann and Wiley 1977). However, if one accepts a most parsimonious topology, then there are situations in which the only statistically acceptable relationship between sister taxa will be one of ancestor and descendant. The assertion that sister taxa must share the same time of origin assumes not only a particular pattern of evolution but also that ancestral species are not included in the analysis. In the resulting cladograms, ancestors and descendants (if there are any such pairs in the species analyzed) will appear, rather, as sister taxa. Yet it is impossible to construct an evolutionary scenario in which ancestors do not precede their descendants. Figure 4.11 shows a hypothetical case in which a fossil species appeared significantly earlier than its extant sister taxon (see Strauss and Sadler 1989, and Marshall 1990 and this volume). The only sister taxon hypothesis that cannot be rejected here is that of ancestor and descendant (Paul 1992).

Finally, it is important to note the distinction between ancestral species and ancestral populations. Although no hypothesis about speciation modes predicts that ancestral *species* should necessarily be rarer in the fossil record than their descendants, only models of gradual anagenesis predict that ancestral *populations* should be sampled relatively frequently. Other models, such as peripheral isolation, predict that ancestral populations (i.e., the peripheral isolates) are restricted in space and time, and are thus poor candidates for being sampled from the fossil record (Eldredge and Gould 1972).

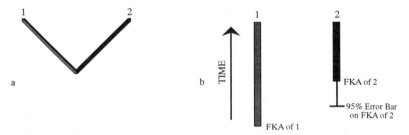

Figure 4.11   Confidence limits on two species demonstrating that it is highly improbable that they share the same time of origin. If it is hypothesized that they are sister taxa, then the only probable relationship is that species 1 in (a) is the ancestor to species 2. "FKA" = first known appearance.

Analysis of Phylogenies of Fossil Taxa

We present two examples of phylogenies of fossil taxa using real clades. The first uses two Neogene clades of foraminifera, the Globigerinidae and the Globorotaliidae. The phylogenetic hypotheses were the work of Kennett and Srinivasan (1983), although the evolutionary tree was taken from Stanley, Wetmore, and Kennett (1988). The second example uses the Ordovician members of a gastropod family, the Lophospiridae. A cladistic analysis of that taxon is presented in a paper by one of us (Wagner 1995).

There are two important differences between the two analyses. First, stratophenetic techniques (e.g., Gingerich 1979) were used to reconstruct the phylogenies of the forams whereas cladistic methods were used for that of the gastropods. Second, forams have extremely complete fossil records, and their sampling is as close to that of computer simulations as is possible from the fossil record; early Paleozoic gastropods, on the other hand, have a much sparser record. If the fossil record of forams is not good enough to execute the method suggested here, then it likely cannot be done for any taxa. Conversely, if speciation patterns within Ordovician gastropods can be examined with phylogenies, then it certainly should be possible to do the same within taxa with denser fossil records (e.g., post-Paleozoic molluscs or Paleozoic brachiopods).

We subjected the phylogenies to two types of analysis. First, we examined whether any of the three basic speciation patterns (cladogenesis, anagenesis, or bifurcation) appear in significantly greater proportions than expected. Second, we examined whether there is any association between the temporal and geographic ranges of a species, its likelihood of leaving descendants (or how many), as well as the type of speciation that a lineage is most likely to undergo. We used simple binomial tests to examine the distributions of speciation patterns, and Mann-Whitney and Kruskal-Wallis tests to examine the possible associations between plesiomorphic species and temporal or geographic ranges. The latter two tests are non-parametric analogs of analysis of variance (ANOVA), which are appropriate given that normal distributions could not be assumed for any of the variables involved. SYSTAT® for Macintosh computers (Feldman et al. 1988) was used to perform non-parametric tests.

Cenozoic Globigerinids

Because most workers use cladograms to present phylogenetic hypotheses, we have converted the evolutionary trees presented by Stanley, Wetmore, and Kennett (1988) into cladistic topologies for this discussion (figure 4.12). The resulting phylogeny reveals 40 speciation events. Five of these appear to be anagenetic, whereas 35 are cladogenetic. No bifurcation appears. A binomial test rejects a hypothesis that cladogenesis and anagenesis should occur in roughly equal proportions. In order to produce probability values greater than 0.05 from the binomial test, it must be postulated that anagenesis accounted for only 28% of the speciation events.

To examine the possible association between likelihood of speciation and temporal and geographic ranges, we divided the species into three classes: those with no likely descendants, those with one, and those with two or more (most non-parametric tests become increasingly less useful with very small counts in categories; because very few species had more than two descendants, we lumped all species with two or more descendants into the same category).

Figure 4.13a shows the temporal ranges of globigerinid species optimized onto the cladogram. Note that the ranges of plesiomorphic species are given at the nodes, whereas the ranges of apomorphic species are given at the ends of the cladogram. Table 4.2 gives the means and standard deviations of the temporal ranges for each group. A significant association exists between lengthier durations of species and their likelihood of leaving descendants ($H = 7.97$, $p = 0.0186$). However, this association applied only to species that gave rise to descendants via cladogenesis. If anagenetic ancestors are excluded, the association is much more significant (table 4.3). Table 4.4 shows that anagenetic ancestors tend to have shorter temporal ranges than have other species, although this tendency is not significant.

To examine the association between speciation and geographic ranges, we used the number of biogeographic units occupied by a species as a proxy for geographic range (according to table 2 in Stanley, Wetmore, and Kennett 1988). Figure 4.13b displays the relationship between plesiomorphy and geography, revealing similar patterns to those of temporal ranges. There is a strong positive association between leaving descendants and possessing wide geographic ranges (Kruskal-Wallis $H = 8.11$,

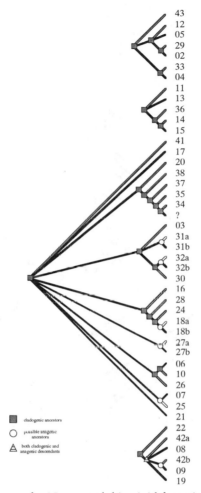

Figure 4.12   Cladogram for Neogene globigerinid foraminifera (derived from Stanley, Wetmore, and Kennett 1988). Species are numbered as in Stanley, Wetmore, and Kennett 1988, with letters denoting different morphospecies within anagenetic lineages.

p = 0.0173), especially via cladogenesis (Kruskal-Wallis H = 9.76, p = 0.0076). However, there was no association between the number of cladogenetic descendants and either temporal or geographic ranges of ancestral species (Mann-Whitney test p = 0.643).

We conducted a similar analysis for another Neogene foram clade, the

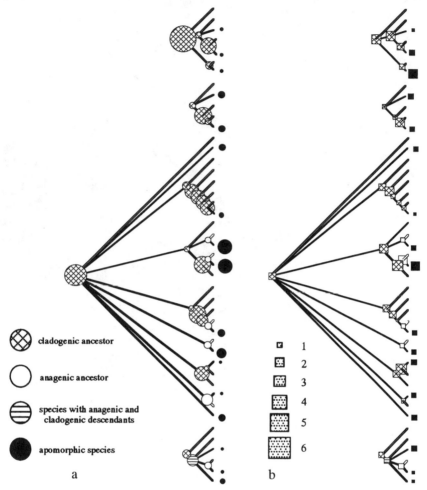

Figure 4.13 Cladogram of globigerinid species (same as 4.12, but with additional information depicting (a) the relative temporal ranges and (b) the relative number of biogeographic units occupied by each species). Geographic ranges are given at the nodes. Black circles or squares at the ends of the cladograms signify species with no known descendants (apomorphic species). The diameters and widths of these circles and squares are proportional to the temporal and geographic range.

Table 4.2

*Association Between Number of Descendants and Temporal Range
Among Globigerinid Foraminifera*

Number Descendants	Number Species	Temporal Range (average My)
0	18	5.39 ± 4.21
1	22	10.25 ± 6.99
2+	6	18.02 ± 11.00

SOURCE: Temporal ranges taken from Stanley, Wetmore, and Kennett 1988.
NOTE: Results of Kruskal-Wallis test are $H = 7.97$, $p = 0.0186$.

Table 4.3

*Association Between Number of Cladogenetic Descendants
and Temporal Range among Globigerinids*

Number Descendants	Number Species	Temporal Range (average My)
0	22	4.85 ± 4.09
1	17	12.60 ± 6.08
2+	6	15.30 ± 11.88

NOTE: Association between cladogenesis and geographic range is $H = 9.76$, $p = 0.0076$; results of Kruskal Wallis test are $H = 14.53$, $p = 0.0007$.

Table 4.4

*Association Between Number of Anagenetic Descendants and
Geographic Range among Globigerinids*

Number Descendants via Anagenesis	Number Species	Temporal Range (average My)
0	40	9.73 ± 7.59
1	5	4.70 ± 5.61

NOTE: Association between geographic range and anagenesis is $Z = -0.46$, $p = 0.6425$; results of Mann-Whitney test are $U = 88$; $p = 0.0577$.

globorotaliids. To conserve space, the results are not presented here, but they were similar to those found for globigerinids: cladogenesis is significantly more common than anagenesis, a positive association exists between having long temporal ranges and leaving cladogenetic descendants, and no such association exists for anagenetic ancestors. However, the association between ancestral species and geographic ranges was not significant, even for cladogenetic ancestors.

There are no previous hypotheses about predominant modes of speciation in globigerinids against which to compare our results. However, such hypotheses should predict the following: (1) a heavy predominance of cladogenesis rather than anagenesis; (2) little or no bifurcation; (3) an increased likelihood of cladogenetic speciation if factors encourage longer temporal ranges (or realized longer temporal ranges); (4) a similar association between cladogenetic speciation and wider geographic ranges; and (5) anagenetic "morphospecies" possessing shorter durations than species prone to cladogenesis. Hypotheses for the globorotaliids should be similar, save for point 4, as there is no association between geographic range and leaving descendants.

Ordovician Lophospirids

Unlike the foram data, all data for the Lophospiridae are inferred from a cladogram (figure 4.14). As detailed elsewhere (Wagner 1995), this is not the most parsimonious overall cladogram, but it is the most parsimonious one that cannot be rejected by stratigraphic data. There have been no previous phylogenetic analyses for this family, although workers have postulated multiple descendants from a few wide-ranging species (e.g., Ulrich and Scofield 1897; Tofel and Bretsky 1987).

Our preferred cladogram for lophospirids (figure 4.14) is rife with polytomies. Of the eleven polytomies, only two do not include plesiomorphic species. Thus, nine may represent hard polytomies. Similarly, of the nine pectinate topologies, eight include plesiomorphic species. All 23 apomorphic species postdate apomorphic sister lineages in the fossil record, although this is statistically significant for only 11 cases. However, in those cases, the only tenable sister-taxon hypothesis is that of ancestor and descendant.

By incorporating data as to stratigraphic ranges, we found that only six of 42 implied speciation events could have been anagenetic and that only one could represent bifurcation. A binomial test strongly rejects the idea that anagenesis and cladogenesis were equally likely among lophospirids ($p = 1.419 \times 10^{-6}$). It must be postulated that anagenesis accounts for only 27% of speciations to produce probability values greater than 0.05.

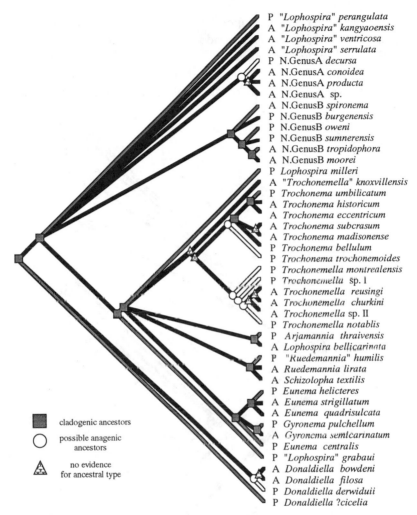

P   *"Lophospira" perangulata*
A   *"Lophospira" kangyaoensis*
A   *"Lophospira" ventricosa*
A   *"Lophospira" serrulata*
P   N.GenusA *decursa*
A   N.GenusA *conoidea*
A   N.GenusA *producta*
A   N.GenusA sp.
A   N.GenusB *spironema*
P   N.GenusB *burgenensis*
P   N.GenusB *oweni*
P   N.GenusB *sumnerensis*
A   N.GenusB *tropidophora*
A   N.GenusB *moorei*
P   *Lophospira milleri*
A   *"Trochonemella" knoxvillensis*
P   *Trochonema umbilicatum*
A   *Trochonema historicum*
A   *Trochonema eccentricum*
A   *Trochonema subcrasum*
A   *Trochonema madisonense*
P   *Trochonema bellulum*
P   *Trochonema trochonemoides*
P   *Trochonemella montrealensis*
P   *Trochonemella* sp. I
A   *Trochonemella reusingi*
A   *Trochonemella churkini*
A   *Trochonemella* sp. II
P   *Trochonemella notablis*
P   *Arjamannia thraivensis*
A   *Lophospira bellicarinata*
P   *"Ruedemannia" humilis*
A   *Ruedemannia lirata*
A   *Schizolopha textilis*
P   *Eunema helicteres*
A   *Eunema strigillatum*
A   *Eunema quadrisulcata*
P   *Gyronema pulchellum*
A   *Gyronema semicarinatum*
P   *Eunema centralis*
P   *"Lophospira" grabaui*
A   *Donaldiella bowdeni*
A   *Donaldiella filosa*
P   *Donaldiella derwiduii*
P   *Donaldiella ?cicelia*

◼   cladogenic ancestors

◯   possible anagenic
      ancestors

△   no evidence
      for ancestral type

Figure 4.14   Cladogram for Ordovician Lophospiridae, including information
on probable patterns of cladogenesis and possible patterns of anagenesis. P denotes ancestral, or plesiomorphic, species; A denotes apomorphic species.

We can now explore the possible association between likely ancestors
and stratigraphic and geographic ranges. As with the forams, we divided
lophospirid species into three groups for the Kruskal-Wallis test: those
with no likely descendants, those with one possible descendant, and those
with two or more descendants. Given the apparent predominance of clad-

ogenesis, the same tests for temporal or geographic association were performed. Figure 4.15 presents the cladogram of the lophospirids with relative temporal ranges optimized onto it in the same manner as for the foram cladogram. Temporal ranges are derived from Harland et al. 1990; however, the same results were obtained using the time scales of Harland

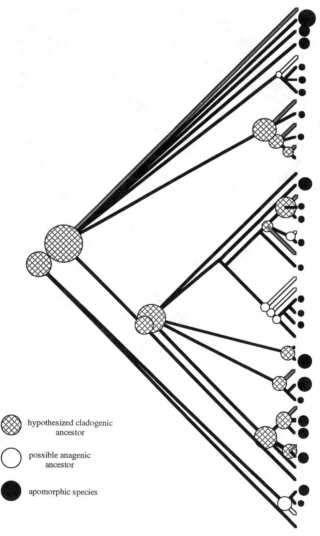

Figure 4.15   Cladogram for Ordovician Lophospiridae (same as figure 4.14), depicting relative temporal ranges. The diameter of each circle is proportional to the temporal range. See the appendix for actual temporal data.

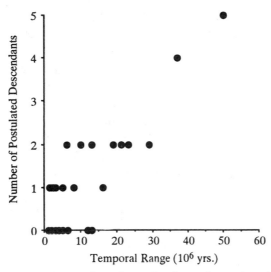

Figure 4.16 The number of putative descendants plotted against temporal range for Ordovician lophospirid species. Temporal ranges derived from Harland et al. 1990.

Table 4.5

Association Between Temporal Ranges of Plesiomorphic Lophospirid Gastropods and the Number of Lineages or Clades to Which They Are Primitive

Number Descendants	Number Species	Temporal Range (average My)
0	25	2.98 ± 3.29
1	11	5.41 ± 4.33
2+	10	21.90 ± 15.31

NOTE: Temporal ranges are determined from Harland et al. 1990. Results of Kruskal-Wallis test are H = 20.63, p < 0.0001. A significant difference also exists between apomorphic species and all plesiomorphic ones (Mann-Whitney: U = 84.33, p = .0001).

et al. 1982 and Ross et al. 1982. Figure 4.16 plots the number of hypothesized descendants against the observed temporal range of the hypothesized ancestor. Table 4.5 compares the association between the temporal ranges of plesiomorphic species and the number of clades or lineages to which they are primitive. Again, a significant association exists between plesiomorphic species and wider temporal ranges (Kruskal-Wallis H = 20.63, p < 0.0001). In general, plesiomorphic species have longer temporal ranges than apomorphic ones (Mann-Whitney test $p < 0.0001$). However, among plesiomorphic species, possibly anagenetic ancestors have significantly shorter temporal durations (Mann-Whitney test $p = 0.01$; table 4.6) and are not distinguishable from apomorphic species.

Figure 4.17 optimizes geographic ranges onto the cladogram. These were approximated from the known distributions of lophospirid horizons and the paleogeographic reconstructions of Scotese (1989). Unlike Neogene forams, the association between plesiomorphic species and wider geographic ranges is significant (tables 4.7 and 4.8). Among plesiomorphic species, a significant association also exists between possible cases of anagenesis and restricted geographic ranges.

The relationship between plesiomorphy and both temporal and geographic ranges in the lophospirids we analyzed suggests that: (1) longer temporal ranges promoted both wider geographic ranges and increased likelihoods of cladogenesis; (2) wider geographic ranges promoted longer temporal ranges and increased the likelihood of cladogenesis; or (3) some factor (or set of factors) encouraging longer durations and greater geographic ranges of species also encouraged cladogenesis. Nearly opposite statements can be made about the relationships between anagenesis and both temporal and geographic ranges. Therefore, any hypothesis about dominant modes of speciation among Ordovician lophospirids should be able to predict the following: (1) cladogenesis is significantly more common than anagenesis; (2) species with longer temporal and geographic ranges are more likely to leave descendants via cladogenesis *or* the factors contributing to wider temporal and geographic ranges also contribute to the likelihood of cladogenetic evolution; (3) if anagenesis occurs, it only applies to species with restricted temporal and geographic ranges; and (4) bifurcation accounts for a negligible amount of speciation.

Table 4.6

Association Between Temporal Ranges of Lophospirid Gastropods and Hypothesized Cladogenetic and Anagenetic Ancestors

Ancestral Type	Number Species	Temporal Range (average My)
Cladogenetic	15	17.23 ± 15.56
Anagenetic	6	4.33 ± 2.86

NOTE: Results of Mann-Whitney test are $U = 75$, $p = 0.01$.

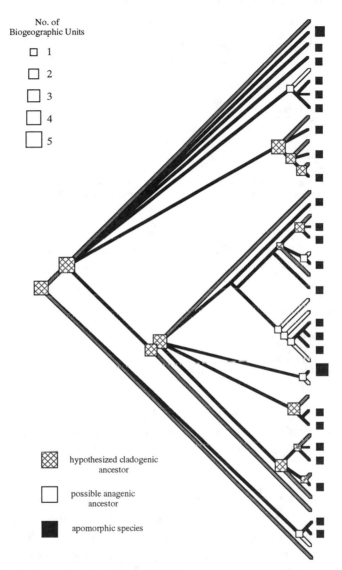

Figure 4.17 Lophospirid cladogram depicting relative geographic ranges, as inferred from Scotese 1989. Key for geographic ranges is 1: < 10⁴ km²; 2: 10⁴ km² to 10⁵ km²; 3: 10⁵ km² to 10⁶ km²; 4: 10⁶ km² to 10⁷ km²; 5: > 10⁷ km². See the appendix.

Table 4.7

Association Between Plesiomorphic Species and
Geographic Ranges of Lophospirids

Number Descendants	Number Species	Number Geographic Units (average)
0	25	1
1	11	1–2
2+	10	3–4

NOTE: Results of Kruskal-Wallis test are H = 10.60, p = 0.005. A significant difference does not, however, exist between apomorphic species and all plesiomorphic ones (Mann-Whitney test p = 0.713).

Table 4.8

Association Between Geographic Ranges and Hypothesized Cladogenetic
and Anagenetic Ancestors of Lophospirids

Ancestral Type	Number Species	Number Geographic Units (average)
Cladogenetic	15	3–4
Anagenetic	6	1

NOTE: Results of Mann-Whitney test are p = 0.0195.

Finally, it bears noting that if the phylogeny of a taxon such as the lophospirids can limit speciation hypotheses, then the methods outlined here should be applicable to a wide range of fossil taxa.

Comparisons with Previous Work

Other workers have suggested associations between temporal/geographic ranges and patterns of speciation (e.g., Scheltema 1978; Jablonski and Lutz 1983; Jablonski 1986). For example, Scheltema (1978) predicted that species of prosobranch gastropods with planktotrophic larvae generally have greater temporal and geographic ranges, and are prone to cladogenesis. Scheltema also predicted that species with nonplanktotrophic larvae are restricted in time and space, and are more prone to anagenesis. The data presented here corroborate part of this hypothesis: the association between patterns of speciation and temporal/geographic ranges. However, they cannot address the association between differences in fundamental population dynamics and speciation patterns, as there is no evidence for variation in basic population dynamics within the taxa analyzed here.

Other workers have focused primarily on rates of speciation. For example, Jablonski (1986) found that species with nonplanktotrophic larvae were more likely to undergo anagenesis and were prone to have faster speciation rates. We have not addressed the question of speciation rates, but the model presented here could be modified to examine differential rates of evolution.

Implications for Phylogenetic Systematics and Comparative Biology

Because bifurcation has been assumed to be the dominant pattern of speciation, many polytomies have been dismissed almost despairingly as unresolved. However, in both examples considered here, little (if any) bifurcation was uncovered. Imposing a bifurcating model on the data would result in gross misinterpretations of the clades' evolutionary histories. A key point of this paper is that patterns of speciation should be tested by phylogenies—not simply assumed when generating and deciphering cladograms. Thus, many polytomies actually may reflect evolutionary patterns. This represents a testable possibility that should be explored in all cases.

A similar logical problem affects recent methodological advances in comparative biology. The main point of Felsenstein 1985, that species cannot be treated as independent entities, is well taken. However, recent phylogenetic models of comparative biology (e.g., Maddison 1990; Harvey and Pagel 1991; Martins and Garland 1991) generally assume that evolution during speciation is no different from evolution within species. Whenever possible this assumption should be tested by phylogenies rather than assumed when those phylogenies are used in comparative biology. Such testing probably is not possible for studies using taxa with poor fossil records or concerning traits that do not fossilize (e.g., Coddington 1988; Gittleman 1986a, 1986b; Losos 1990, 1992; Lynch 1991). However, given a phylogenetic hypothesis, comparative biological studies using morphologic data can test whether intraspecific evolution is the same as interspecific evolution. The one such study to date disputes this assumption (Cheetham 1986, 1987; Cheetham and Jackson, this volume;

but also see Marshall, this volume). More such studies are needed before generalizations can be made, but in all cases paleobiologists should test and justify the assumptions of the particular comparative method that they use.

Applicability and Limitations

Only species-level phylogenies should be analyzed in the manner set forth in this paper. Genus or family level phylogenies omit too many speciation events and cannot exactly describe the timing of the included events. Moreover, it is possible that morphologic changes great enough to be recognized as new genera or families do not have the same type of distribution within a phylogeny as do standard speciation events. We are not claiming that this possible distinction between specific and supraspecific evolution is necessarily true, but it is something that should be tested using species-level phylogenies. Additionally, clades should be well sampled at the species level. Good sampling not only increases the robustness of a phylogenetic hypothesis (Paul 1982, 1985; Allmon 1989) but also diminishes the tricky problems of gaps that accompany analyses of evolutionary relationships above the species level.

In the introduction, we mentioned that phylogenies can corroborate or falsify hypotheses about speciation modes. However, the methods discussed here cannot demonstrate whether a particular hypothesis is correct. More than one mode of speciation may be consistent with a particular phylogenetic result, or the fossil record may not allow discrimination between competing hypotheses. The modes of speciation producing cladogenesis, anagenesis, and perhaps even bifurcation can be either punctuated or gradual (MacLeod 1991). Well-documented phylogenies can show stasis within lineages (in a loose sense, not *sensu* Bookstein 1987, 1988) if MBL-type patterns of evolution are demonstrated. However, a phylogeny cannot demonstrate whether the cladogenetic speciation was gradual or punctuated.

Although phylogenies and cladograms cannot reveal modes of speciation, there is a danger that they create the impression of only punctuated change (Gayon 1990). Traditional ("alpha") systematics relies on a combination of both discrete, meristic characters (i.e., things one can count) and continuously variable characters (i.e., shapes and sizes). In contrast, cladistic analyses require that both types of characters be described as if they were discrete. This is not because systematists are ignorant of the importance of continuously variable characters, but simply because reliable and repeatable coding of differences in shape and size is a very difficult problem (e.g., see Thorpe 1984; Archie 1985; Goldman 1988).

The potential bias toward perceiving punctuated change may be exacerbated by the fact that a cladogram contains no information regarding the timing of character state changes. A hypothetical example is given in figure 4.18. If specimens are sampled and coded from three points along a gradually evolving lineage (figure 4.18a), the correct cladogram (figure

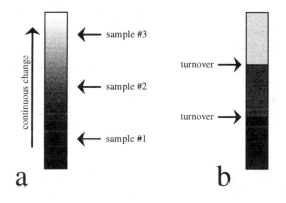

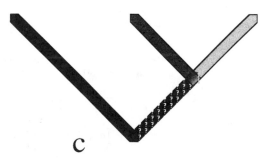

Figure 4.18 An example of how gradual and punctuated evolution (*sensu* MacLeod 1991) cannot be distinguished on a cladogram. A cladistic analysis of specimens from the three points along a gradually anagenetic lineage (a) should lead to the cladogram shown in c. This cladogram could inadvertently give the impression that punctuated anagenetic change occurred (b), despite lack of supporting evidence.

4.18c) will link the later specimens as sister taxa using qualitative states of continuously evolving characters. A literal reading of this may give the impression that punctuated anagenesis occurred (figure 4.18b), but this would be purely an artifact of the methods of phylogenetic analysis.

Hypotheses about modes of speciation make predictions about patterns of speciation, which in turn make predictions about patterns within phylogenies. Therefore, phylogenetic patterns offer a potential test of speciation hypotheses that can corroborate or disprove (but, alas, not prove) those hypotheses. Two examples we have drawn from the fossil record indicate that testing of speciation hypotheses can be done not only for taxa with excellent fossil records but also for those with more average records. Thus, even if the fossil record of a clade leaves no direct evidence about common modes of speciation, the clade's phylogenetic patterns can be used to limit the appropriate hypotheses.

Appendix

Temporal and Geographic Ranges of Lophospirid Species and the Number of Lineages to Which They Were Plesiomorphic

Temporal ranges are derived from Harland et al. 1990. Geographic ranges were approximated using the reconstructions in Scotese 1989. See the caption of figure 4.17 for the key to the geographic ranges signified by the numbers that appear in that figure and here. *A.* = *Arjamannia*; *D.* = *Donaldiella*; *E.* = *Eunema*; *G.* = *Gyronema*; *L.* = *Lophospira*; *R.* = *Ruedemannia*; *S.* = *Schizolopha*; *Ta.* = *Trochonema*; *Tl.* = *Trochonemella*.

Species	Temporal Range (My)	Geographic Range	Possible Descendants
D. ?cicelia	29	4	2
"*L.*" *perangulata*	50	5	5
"*L.*" *kangyaoensis*	12	2	0
"*L.*" *grabaui*	16	3	1
L. milleri	37	5	4
D. derwiduii	8	1	1
"*Tl.*" *knoxvillensis*	13	1	0
E. centralis	21	2	2
Ta. trochonemoides	1.5	1	1
Ta. bellulum	6	1	2
Ta. eccentricum	4	1	0

Tl. montrealensis	3	1	1
E. helicteres	13	1	2
"L." ventricosa	5	1	0
"L." serrulata	5	1	0
Tl. sp. I	2.5	1	1
Ta. umbilicatum	19	2	2
E. strigillatum	5	1	0
Tl. notabilis	2.5	1	1
G. pulchellum	8	1	1
G. semicarinatum	2	1	0
N. Gen. B burgenensis	23	4	2
N. Gen. B spironema	1	2	0
N. Gen. B oweni	10	2	2
N. Gen. B sumnerensis	2	1	0
Ta. historicum	3	1	0
Ta. subcrasum	8	1	1
"R." humilis	6	2	2
N. Gen. A sp.	1	1	0
N. Gen. A decursa	2	1	1
N. Gen. A conoidea	1	1	0
N. Gen. A producta	1	1	0
S. textilis	1	1	0
D. bowdeni	6.5	1	0
D. filosa	1	1	0
N. Gen. B tropidophora	5	2	1
R. lirata	3	1	0
L. aff. L. serrulata	1	1	0
Tl. churkini	1	1	0
Tl. reusingi	1	1	0
L. bellicarinata	3	1	1
E. quadrisulcata	3	1	0
A. thraivensis	4	3	0
N. Gen. B moorei	1	1	0
Ta. madisonense	2	1	0

Acknowledgments   P. W. would like to thank the editors of this volume first for the invitation to present a paper at the 1992 G.S.A. symposium they organized, and second for the invitation to be a contributor (with coauthor Douglas Erwin). This paper originated from a class project by P. W. concerning biases in the fossil record, which was undertaken for David Raup's "Evolutionary Paleobiology" course. Thanks are owed to him for making the original MBL notes available. Both authors would like to thank the following people who critiqued various portions and iterations of the manuscript or provided valuable discussion: J. Alroy, G. Eble, D. Jablonski, E. E. LeClair, R. Lupia, D. M. Raup, and J. J. Sepkoski, Jr.

References

Allmon, W. D. 1989. Paleontological completeness of the record of lower Tertiary Mollusks, U.S. Gulf and Atlantic coastal plains: Implication for phylogenetic studies. *Historical Biology* 3:141–158.

Alroy, J. 1992. Conjunction among taxonomic distributions and the Miocene mammalian biochronology of the Great Plains. *Paleobiology* 18:326–343.

Archie, J. W. 1985. Methods for coding variable morphological features for numerical taxonomic analysis. *Systematic Zoology* 34:326–345.

Bookstein, F. L. 1987. Random walk and the existence of evolutionary rates. *Paleobiology* 13:446–464.

———. 1988. Random walk and the biometrics of morphologic characters. *Evolutionary Biology* 23:369–398.

Boucot, A. J. 1978. Community evolution and rates of cladogenesis. *Evolutionary Biology* 11:545–655.

Brooks, D. R. and D. A. McLennan. 1991. *Phylogeny, ecology and behavior: A research program in comparative biology.* Chicago: University of Chicago Press.

———. 1986. Tempo of evolution in a Neogene bryozoan: Rates of morphologic change within and across species boundaries. *Paleobiology* 12:190–202.

Cheetham, A. H. 1987. Tempo of evolution in a Neogene bryozoan: Are trends in single morphologic characters misleading? *Paleobiology* 13:286–296.

Coddington, J. A. 1988. Cladistic tests of adaptational hypotheses. *Cladistics* 4:3–22.

Cracraft, J. 1981. Pattern and process in paleobiology: The role of cladistics in systematic paleontology. *Paleobiology* 7:456–468.

———. 1989. Speciation and its ontology: The empirical consequences of alternative species concepts for understanding patterns and processes of differentiation. In D. Otte and J. A. Endler, eds., *Speciation and Its Consequences,* pp. 28–59. Sunderland, Mass.: Sinauer.

Darwin, C. 1859. *The Origin of Species.* New York: Random House.

Eldredge, N. 1971. The allopatric model and phylogeny in Paleozoic invertebrates. *Evolution* 25:156–167.

Eldredge, N. and S. J. Gould. 1972. Punctuated equilibria: An alternative to phyletic gradualism. In T. J. M. Schopf, ed., *Models in Paleobiology,* pp. 82–115. San Francisco: Freeman.

Engelmann, G. F. and E. O. Wiley. 1977. The place of ancestor-descendant relationships in phylogeny reconstruction. *Systematic Zoology* 26:1–11.

Feldman, D., J. Gagne, R. Hofmann, and J. Simpson. 1988. *Statview™ SE + Graphics 1.02.* Berkeley: Abacus Concepts.

Felsenstein, J. 1985. Phylogenies and the comparative method. *American Naturalist* 125:1–15.

Gayon, J. 1990. Critics and criticisms of the modern synthesis—the viewpoint of a philosopher. *Evolutionary Biology* 24:1–49.

Gingerich, P. D. 1976. Paleontology and phylogeny: patterns of evolution at the species level in early Tertiary mammals. *American Journal of Science* 276:1–28.

———. 1979. The stratophenetic approach to phylogeny reconstruction in vertebrate paleontology. In J. Cracraft and N. Eldredge, eds., *Phylogenetic Analysis and Paleontology*. pp. 41–77. New York: Columbia University Press.

———. 1985. Species in the fossil record: Concepts, trends, and transitions. *Paleobiology* 11:27–42.

Gittleman, J. L. 1986a. Carnivore brain size, behavioral ecology, and phylogeny. *Journal of Mammalogy* 67:23–36.

———. 1986b. Carnivore life history patterns: Allometric, phylogenetic, and ecological associations. *American Naturalist* 127:744–771.

Goldman, N. 1988. Methods for discrete coding of morphological characters for numerical analysis. *Cladistics* 4:59–71.

Guyer, C. and J. B. Slowinski. 1991. Comparisons of observed phylogenetic topologies with null expectations among three monophyletic lineages. *Evolution* 45:340–350.

Harland, W. B., R. L. Armstrong, A. V. Cox, L. E. Craig, A. G. Smith, and D. G. Smith 1990. *A Geologic Time Scale 1989*. Cambridge: Cambridge University Press.

Harland, W. B., A. V. Cox, P. G. Llewellyn, C. A. G. Pickton, A. G. Smith, and R. Walters. 1982. *A Geologic Time Scale*. Cambridge: Cambridge University Press.

Harvey, P. H. and M. D. Pagel 1991. *The Comparative Method in Evolutionary Biology*. Oxford: Oxford Press.

Heard, S. B. 1992. Patterns in tree balance among cladistic, phenetic, and randomly generated phylogenetic trees. *Evolution* 46:1818–1826.

Hennig, W. 1966. *Phylogenetic Systematics*. Urbana: University of Illinois Press.

Jablonski, D. 1986. Larval ecology and macroevolution in marine invertebrates. *Bulletin of Marine Science* 39:565–587.

Jablonski, D. and R. A. Lutz. 1983. Larval ecology of marine benthic invertebrates: Paleobiological implications. *Biological Review* 58:21–89.

Kennett, J. P. and M. S. Srinivasan. 1983. *Neogene Planktonic Foraminifera—a Phylogenetic Atlas*. Stroudsburg, Penn.: Hutchinson Ross.

Lande, R. 1980. Genetic variation and phenotypic evolution during peripheral isolation. *American Naturalist* 116:463–479.

———. 1982. Rapid origin of sexual isolation and character divergence in a cline. *Evolution* 36:213–223.

———. 1986. The dynamics of peak shifts and the pattern of morphologic evolution. *Paleobiology* 12:343–354.

Larson, A. 1989. The relationship between speciation and morphologic evolution. In D. Otte and J. A. Endler, eds., *Speciation and Its Consequences,* pp. 579–598. Sunderland, Mass.: Sinauer.

Lorenzen, S. and J. Sieg. 1991. PHYLIP, PAUP, and HENNIG86—how reliable are computer parsimony programs used in systematics? *Zeitschrift fur Zoologisches Systematische Evolution* 22:234–263.

Losos, J. B. 1990. A phylogenetic analysis of character displacement in Caribbean *Anolis* lizards. *Evolution* 44:558–569.

———. 1992. The evolution of convergent structure in Caribbean *Anolis* communities. *Systematic Biology* 41:403–420.

Lynch, J. D. 1989. The gauge of speciation: On the frequencies of modes of speciation. In D. Otte and J. A. Endler, eds., *Speciation and Its Consequences*, pp. 527–553. Sunderland, Mass.: Sinauer.

Lynch, M. 1991. Methods for the analysis of comparative data in evolutionary biology. *Evolution* 45:1065–1080.

MacLeod, N. 1991. Punctuated anagenesis and the importance of stratigraphy to paleobiology. *Paleobiology* 17:167–188.

Maddison, W. P. 1989. Reconstructing character evolution on polytomous cladograms. *Cladistics* 5:365–377.

———. 1990. A method for testing the correlated evolution of two binary characters: Are gains or losses concentrated on certain branches of a phylogenetic tree? *Evolution* 44:539–557.

Maddison, W. P. and M. Slatkin. 1991. Null models for the number of evolutionary steps in a character on a phylogenetic tree. *Evolution* 45:1184–1197.

Marshall, C. R. 1990. Confidence intervals on stratigraphic ranges. *Paleobiology* 16:1–10.

Martins, E. P. and T. Garland Jr. 1991. Phylogenetic analysis of the correlated evolution of continuous characters: A simulation study. *Evolution* 45:534–557.

Maynard Smith, J. 1966. Sympatric Speciation. *American Naturalist* 100:637–650.

Mayr, E. 1963. *Animal Species and Evolution.* Cambridge: Harvard University Press.

———. 1982. Processes of speciation in animals. In C. Barigozzi, ed., *Mechanisms of Speciation*, pp. 1–19. New York: Alan R. Liss Press.

Nixon, K. C. and Q. D. Wheeler. 1992. Extinction and the origin of species. In M. J. Novacek and Q. D. Wheeler, eds., *Extinction and Phylogeny*, pp. 119–143. New York: Columbia University Press.

Panchen, A. L. 1992. *Classification, Evolution and the Nature of Biology.* Cambridge, U.K.: Cambridge University Press.

Paul, C. R. C. 1982. The adequacy of the fossil record. In K. A. Joysey and A. E. Friday, eds., *Problems of Phylogenetic Reconstruction*, pp. 75–117. London: Academic Press.

———. 1985. The adequacy of the fossil record revisited. *Special Papers in Palaeontology* 33:1–16.

———. 1992. The recognition of ancestors. *Historical Biology* 6:239–250.

Provine, W. B. 1989. Founder effects and genetic revolutions in microevolution and speciation: A historical perspective. In L. V. Giddings, K. Y. Ashiro, and

W. W. A. Anderson, eds., *Genetics and the Founder Principle*, pp. 43–60. London: Oxford University Press.

Raup, D. M. 1977. Stochastic models in evolutionary palaeontology. In A. Hallam, ed., *Patterns of Evolution*, pp. 59–78. Amsterdam: Elsevier.

Raup, D. M. and S. J. Gould 1974. Stochastic simulation and evolution of morphology—towards a nomothetic paleontology. *Systematic Zoology* 23:305–322.

Rensch, B. 1959. *Evolution Above the Species Level*. New York: Columbia University Press.

Ross, R. J. J., F. J. Adler, T. W. Amsden, D. Bergstrom, S. M. Bergstrom, C. Carter, M. Churkin, E. A. Cressman, J. R. Derby, J. T. Dutro, R. L. Ethington, S. C. Finney, D. W. Fisher, J. H. Fisher, A. G. Harris, L. F. Hintze, K. B. Ketner, K. L. Kolata, E. Landing, R. B. Neuman, W. C. Sweet, J. Pojeta, Jr., A. W. Potter, E. K. Rader, J. E. Repetski, R. H. Shaver, T. L. Thompson, and G. F. Webers. 1982. *The Ordovician System in the United States*. International Union of Geological Sciences.

Scheltema, R. S. 1978. On the relationship between dispersal of pelagic veliger larvae and the evolution of marine prosobranch gastropods. In B. Battaglia and J. A. Beardmore, eds., *Marine Organisms: Genetics, Ecology and Evolution*, pp. 303–322. New York: Plenum Press.

Scotese, C. R. 1989. Phanerozoic reconstructions: A new look at the assembly of Asia. *Phanerozoic Mapping Project Progress Report* 19:1–10.

Signor, P. W. 1985. Real and apparent trends in species richness through time. In J. W. Valentine, ed., *Phanerozoic Diversity Patterns: Profiles in Macroevolution*, pp. 129–150. Princeton: Princeton University Press.

Slowinski, J. B. and C. Guyer 1989a. Testing null models in questions of evolutionary success. *Systematic Zoology* 38:189–191.

———. 1989b. Testing the stochasticity of patterns of organismal diversity: An improved null model. *American Naturalist* 134:907–921.

Smith, G. R. 1992. Introgression in fishes: Significance for paleontology, cladistics, and evolutionary rates. *Systematic Biology* 41:41–57.

Stanley, S. M., K. L. Wetmore, and J. P. Kennett. 1988. Macroevolutionary differences between two major clades of Neogene planktonic foraminifera. *Paleobiology* 14:235–249.

Strauss, D. and P. M. Sadler. 1989. Classical confidence intervals and Bayesian probability estimates for ends of local taxon ranges. *Mathematical Geology* 21:411–427.

Szalay, F. S. 1977. Ancestors, descendants, sister groups, and testing of phylogenetic hypotheses. *Systematic Zoology* 26:12–18.

Thorpe, R. S. 1984. Coding morphometric characters for constructing distance Wagner networks. *Evolution* 38:244–255.

Tofel, J. E. and P. W. Bretsky. 1987. Middle Ordovician *Lophospira* (Archaeogastropoda) from the upper Mississippi Valley. *Journal of Paleontology* 61:700–723.

Ulrich, E. O. and W. H. Scofield. 1897. The Lower Silurian Gastropoda of Minnesota. *Paleontology of Minnesota* 3:813–1081.

Valentine, J. W. 1989. How good was the fossil record? Clues from the California Pleistocene. *Paleobiology* 15:83–94.

Wagner, P. J. 1995. Stratigraphic tests of cladistic hypotheses, *Paleobiology* 21.

Wake, D. B., K. P. Yanev, and M. M. Frelow. 1989. Sympatry and hybridization in a "ring species": The plethodontid salamander *Ensatina eschscholtzii.* In D. Otte and J. A. Endler, eds., *Speciation and Its Consequences,* pp. 134–157. Sunderland, Mass.: Sinauer.

Wright, S. 1931. Evolution in Mendelian populations. *Genetics* 16:97–159.

———. 1932. The roles of mutation, inbreeding, crossbreeding, and selection in evolution. *Proceedings of the Sixth International Congress of Genetics* 1:356–366.

———. 1982. Character change, speciation, and the higher taxa. *Evolution* 36:427–443.

5

Metapopulations and Disturbance: From Patch Dynamics to Biodiversity Dynamics

Michael L. McKinney and Warren D. Allmon

If you can't measure it, it's just an opinion.
—ROBERT A. HEINLEIN

Despite a vast amount of scientific study during the last twenty years, the validity of punctuated equilibria remains widely disputed (Hoffman 1989; Levinton 1988). Two basic reasons account for this lack of either confirmation or dismissal. First, the tempo and mode of evolution appears to be so variable in both pattern and process that a single model, such as punctuated equilibria, cannot fit. Fossil studies have revealed a wide variety of speciation patterns, ranging from gradual anagenesis and cladogenesis to punctuated equilibria (Gingerich 1985; Levinton 1988; Hoffman 1989; Geary, this volume).

Just because the evolutionary process yields complex patterns, that alone does not account for the long-standing nature of this debate. We contend that the second reason for the continuing uncertainty is the failure of most studies to go beyond documenting patterns, to investigate the dynamics of the process. Most past studies of speciation in the fossil record rely on the same basic strategy: detailed morphological measurements of a lineage in a stratigraphic section that is selected for its relatively high stratigraphic completeness (e.g., Williamson 1987; Sheldon 1987). This has led to debate because the patterns found can be explained in a number of ways. Most important is the past emphasis on stasis as the primary evidence for punctuated equilibria (Gould 1985). Stasis can be explained by stabilizing selection (e.g., Johnson 1982; Charlesworth, Lande, and Slatkin 1982) or by intrinsic constraints (Gould 1989).

This emphasis on stasis has usually been justified by pointing out the difficulty or even impossibility of studying speciation directly in the fossil record: "rapid transitions are mostly recorded by an absence of information" (Gould 1982, p. 86). Stanley (1982) and many others have also made this observation.

Fossil Populations and Metapopulations

This paper outlines specific, testable models by which speciation, and extinction, *can* be directly studied in the fossil record. We argue that moving fossil studies away from a focus on stasis, or equilibria, toward the dynamics of the "punctuations" evident in the record is one of the best ways available to move the debate forward. The need for such a shift is explicit in Hoffman's (1989, p. 123) conclusion that punctuated equilibria has "reached the stage where there is nothing to debate any more." Perhaps more important, we show how the study of speciation and extinction in fossils may be scaled up to account for large-scale biodiversity patterns through geologic time and perhaps in the future.

Our focus is the study of fossil populations. Populations are in many ways the most fundamental unit of both ecological and phylogenetic processes (Eldredge 1992). There is nothing drastically novel about such a focus. Hallam's (1972) classic paper on fossil populations was published in the same pioneering volume that contained the early Eldredge and Gould (1972) paper on punctuated equilibria. Boucot (e.g., 1975) has for years discussed fossil patterns in terms of population processes. The general paleontological interest, however, has been away from population-level studies, in favor of large-scale studies of taxonomic diversity or other characteristics. Where population level studies are most appropriate, such as testing for punctuated versus gradual patterns of speciation, the population approach has been ignored or erroneously conceived.

Gould and Eldredge, for example, have repeatedly emphasized the importance of geographic sampling, but their suggestions have generally been ignored. Many often-discussed high resolution microfossil studies (e.g., Gingerich 1985; MacLeod 1991) come from only a single section. Aside from stratigraphic distortions in this approach (as discussed in MacLeod 1991), the results are of limited value for evolutionary studies because they record change (or nonchange) in only a *local population* or series of local populations, through time. It is true that the huge abundance and pelagic (widespread) nature of these organisms may render the local population a valid proxy for the entire species. But given the great geographic range and many environmental barriers that could reduce gene flow, there must be doubt. Even more dubious are (often widely cited) conclusions drawn from very restricted sampling of terrestrial organisms or benthic organisms with patchy distributions.

The rapidly growing interest in metapopulation theory in ecology provides an excellent opportunity for paleontology to incorporate and test a well-developed body of ideas on how population dynamics through time and space can account for observed fossil patterns of speciation and extinction. Levins (1969) coined the useful term "metapopulation" to describe any species as a number of populations persisting in a dynamic balance between extinction and recolonization (reviews in Hanski 1989; Hanski and Gilpin 1991). Metapopulations are usually distributed as clumps or patches. They have been documented in terrestrial and marine organisms of all types (Hastings and Wolin 1989; Murphy, Freas, and Weiss 1990), including benthic foraminifera (Murray 1991), which will be a focus of this paper. The degree of patchiness varies; a few species have near-continuous population distributions, but others occur in extremely isolated patches. The widely cited work of Pulliam (1988) indicates that many metapopulations have a few large patches of source populations amid many smaller patches of "sink" populations that are sustained by immigration from the sources.

All species probably persist as metapopulations at some spatial scale (Harvey et al. 1993), so the concept has considerable potential for general application. In the present context, the fossil record is the product of more than 3.5 billion years of metapopulation dynamics. Speciation occurred from patch isolation, gene flow restriction, and other events that can be analyzed using metapopulation concepts. These concepts incorporate classical thinking on allopatric speciation, such as the importance of geographic variation in morphology and population structure (Otte and Endler 1989). The metapopulation approach supplements this thinking by permitting a more rigorous description of patchiness and its consequences.

Here we present and test two models of metapopulations in the fossil record. The first model examines fine-scale dynamics by describing how the fossil record of population patches may be directly viewed in local stratigraphic sections. Probably much of the paleontological disinterest in population studies, especially toward analyzing patchiness and geographic variation, has been rooted in the perception that the resolution of the fossil record is too coarse to preserve much evidence. This perception has not been lessened by the rapid growth of taphonomy which, by its nature, focuses on how population information is distorted and omitted in the recording process. Our first model describes why patches may be recorded. We present initial results of data that support this, and discuss some of the many implications of seeing fossil patches in studies of paleocommunities, speciation, and in biostratigraphy.

The second model, and the second part of the paper, examines the coarser-scale dynamics of how metapopulation processes become translated into the biodiversity patterns viewed in the fossil record. This "intermediate disturbance model" describes how such patterns as the correlation of extinction and origination rates, and onshore-offshore origination, may be related to the dynamics of populations. This model is also tested by using an example that is probably representative of many macrofossiliferous strata in its incompleteness. Perhaps the most intriguing implication is the possible continuity of pattern and process across all scales, from patches to higher taxa, because of the self-similar nature of disturbance-driven dynamics. This finding is potentially of great practical importance; it may help predict the future cumulative effects on the biosphere of ongoing small-scale disturbances.

Patch Dynamics in the Fossil Record

Studies in which geographic sampling of fossil species has been attempted usually focused on comparative, statistical analyses of morphological variation. The added dimension of variation in abundance—that is, patchiness—has rarely, if ever, been addressed in any rigorous way in fossils.

A stratigraphic section or core represents a point in space that may record in time the movement of environments and populations across it when deposition of sediment or fossils, respectively, occurs at that point. In classic geological terms, a restricted view of space during an extended passage of time yields the familiar Walther's Law, wherein horizontal spatial changes translate into vertical ones. As shown in figure 5.1, the movement of patches and environments across a point in space is math-

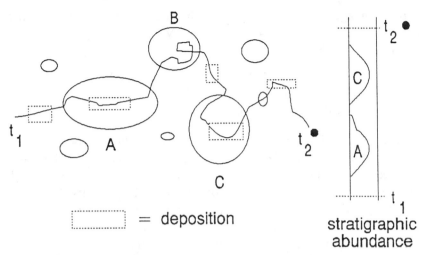

Figure 5.1 Fossil and sediment deposition at a stratigraphic section or core is determined by horizontal migration of patches and facies over the site of deposition. This is analogous to a particle that migrates through the patches (e.g., A, B, C), where the particle is the section or core. More rapid migration reduces the chance that deposition will occur as the patch passes over (e.g., B passes during a time of nondeposition).

ematically equivalent to movement of the point across the patches and environments. In reality, the patches and facies are, through time, moving past the spatially limited sampling point of the stratigraphic section; but the results are the same as if the observer had access to a wider geographic view but was limited to one instant in time. We will sometimes refer to this focus on a very restricted spatial spot as the "particle."

Rate of Patch Migration (Particle Walk)

Two key parameters, the rate and the direction of "particle" movement, will affect what is deposited through time at the point in space (the section) being sampled. Turning first to rate of movement, figure 5.1 shows that faster movement of the particle—that is, rapid patch migration—will tend to reduce the amount of fossil deposition of any given patch by reducing the time over which deposition occurs. Conversely, other things being equal, slower movement of the particle (more time on the patch) permits more time for deposition. We can also consider rate of fossil deposition on the particle as it moves. The rate of particle movement and the rate of deposition can be simultaneously expressed as a ratio, which might be called the "index of patch resolution," or IPR.

$$\text{IPR} = (\text{depositional rate}) / (\text{particle velocity})$$
$$= (\#\text{fossils} / \text{time}) / (\text{m} / \text{time})$$
$$= \#\text{fossils} / \text{m}$$

In these equations particle movement is horizontal distance (m) traveled per unit time. A hypothetical constant depositional rate of 10 fossils per week on a particle that travels a constant velocity of 1 meter per week will deposit 10 fossils per meter. In other words, the patch itself migrates 1 meter per week and deposits 10 fossils during that time at the section. The key message of the IPR is that the higher the value of the ratio, the greater is the patch resolution: as depositional rate increases and as particle movement (patch migration) decreases, smaller and smaller patch sizes can be expected to contribute at least one individual to the fossil record.

We are not suggesting that each and every patch that would be expected in a sample would indeed be there. In the real world, neither depositional rate nor particle velocity is constant, so the resolution will fluctuate with these, and there are many taphonomic biases (discussed later), especially "time-averaging," that prevent such fine, per patch visibility. However, we are suggesting that these processes may not destroy *all* evidence on patchiness in fossil species; rare, patchy species should often be distinguishable from very abundant, more uniformly distributed species. This distinction is feasible because patchiness occurs at many scales, as a hierarchy of clusters of individuals which in turn form clusters of patches, which form larger patches, and so on (Hanski and Gilpin 1991). The key implication is that long-term time-averaged views of patchiness may preserve the finer levels of the hierarchy.

Patch Density: Clues from Depositional Rate

Patch size (area) and patch density (abundance of individuals per unit area) are the two most basic ways to characterize a patch. The main clue on patch density comes from fossil depositional rate: we might expect that the more abundant a living population, the more fossils will be deposited. The simplest assumption is that fossil depositional rate is directly proportional to life abundance. For example, assume that two species coexist on the same 1 m patch and that one species is ten times more abundant than the other. The more abundant species may deposit 100 fossil/m while the other deposits 10/m of particle traveled, leaving 100 and 10 fossils in the section, respectively. Both species have had their 1 m patch recorded, and their differing abundances are also recorded.

The crucial assumption here, that fossil depositional rate is proportional to true abundance, is undoubtedly simplistic. Many biases can occur in the conversion of a biocoenosis into a taphocoenosis (Allison and Briggs 1991; Kidwell and Behrensmeyer 1993). Mixing of layers can lead to time averaging of abundances, changes in sediment (matrix) deposition can alter abundance per unit volume, and destruction of individuals can occur via dissolution, recrystallization, and other processes. Although there is no doubt that such taphonomic processes can dampen and distort the original abundance "signal," there is growing empirical evidence that taphonomic abundance is at least partly correlated with original life abundance. One example is presented by Fursich and Aberhan (1990), who discuss how ranked taphonomic abundance is relatively constant, apparently as a result of long-term stability of conditions. Similarly, Miller (1988) shows that the deposited remains of a marine community preserve many of the fine abundance details of the living community. Koch (1991) reviews two abundance patterns in the record that mimic those of living species: (1) Correlation of local abundance with geographic range is strikingly similar (compare Koch 1991, table 2, with Hengeveld 1992, p. 147). (2) Log-series abundance patterns within communities indicate that most species are rare in both fossil and living communities.

Variable Rates of Particle Velocity and Fossil Deposition

Even if we assume that fossil depositional rates are valid indicators of original population abundance, we must still acknowledge that depositional rates of fossils and sediment are not constant but are episodic. Any approximate correlation between biocoenotic and taphonomic abundance is the result of a lengthy time-averaging process wherein the depositional rates represent some mean of all the depositional episodes and periods of nondeposition (Sadler and Strauss 1990). Similarly, we know that particle (patch) velocity is not constant. More rigorously:

$$\frac{dD/dt}{dH/dt} = IPR \qquad \text{(eq. 1)}$$

where D is number of fossils deposited and H is horizontal distance moved by the particle. If we want to incorporate sedimentation, then:

$$\frac{\left(\dfrac{dD/dt}{dS/dt}\right)}{(dH/dt)} = \frac{\text{volumetric abundance}}{\text{particle velocity}} \qquad \text{(eq. 2)}$$

where S is amount of sediment deposited. If we wish to factor in preservation, we would substitute P for D, based on an assumption of proportionality we discuss later.

Because none of the velocity terms are truly constant, these expressions summarize in explicit terms what we all know: depositional rates of both fossils and sediments are episodic and highly variable, with much time having little or no deposition (Sadler and Strauss 1990). Even if we assume some mean fossil and sediment depositional rate for a particular case, the highly skewed (non-normal) frequency distribution of rates makes the mean of little value.

The problem of variable depositional and migration rates can be approached in a number of ways, using empirical data on depositional rates combined with theoretical models of deposition that assume log-normal and other skewed rate distributions. These complications do not, however, obviate the basic relationship. The variable rates in the top and bottom of the ratio merely suggest that patch resolution will vary through time, as the ratio varies through time. Initial models indicate that surprisingly small patches, as low as 1 m diameter, will often leave at least a few fossil representatives of small organisms, such as benthic foraminifera, provided that depositional rates are relatively high and that patch migration is relatively slow—say, one meter per month.

Patch Size: Clues from Particle Direction

So far, our focus has been on velocity of the particle, or stratigraphic section, as an analytical analog of patch velocity. The main practical use of this, as we have shown, is that it can indicate patch density, that is, fossil abundance of a species on the patch. We turn now to the second basic property of the particle analog, that of particle direction.

A main advantage of viewing the stratigraphic section as a moving particle is that it can be analyzed with the well-developed models of particle diffusion (see Berg 1981 for a review of the biological contexts). For example, a two-dimensional random walk of a particle on the same horizontal stratigraphic surface can be an exact model for what would be recorded vertically through time at a single, unmoving spot if for thousands or even millions of years sea level and other major environmental regional controls are undergoing no long-term directional trends: facies migrate "randomly" back and forth across the sea floor in all directions in response to local, short-term fluctuations in conditions. Where nonrandom directional trends occur over the long term (such as a marine regression that prompts facies migrations seaward over many thousands of years), the process can be accurately modeled as "diffusion with drift" (Berg 1981). In this case, the particle walk still has a strong random component, but with a probabilistic tendency to go in one general direction (set of associated vectors) in the long term. Increasing strength of drift is modeled by increasing the probability that a particle will move in certain directions.

Figure 5.2 shows an example of a computer-generated 2-D random walk. It illustrates one of the deepest insights from particle diffusion theory: a randomly moving particle must often travel for a very long time (or absolute distance) to attain significant displacement in a single direction. More rigorously, the mean displacement of a particle from its origin on a two-dimensional random walk is only (2 × total distance traveled)$^{1/2}$. On average, the randomly moving particle will move away from its current position only as the square root of total distance traveled, whether the latter is measured in steps or other unit distance. In the example of figure 5.2, the particle was displaced a mere 196 steps from origin after

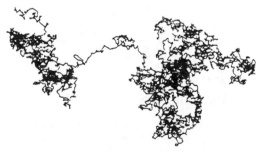

Figure 5.2   A two-dimensional random walk of n = 18,050 steps. Repeatedly traversed areas are more densely black; net horizontal displacement, from upper left to rightmost point, was 196 steps. Predicted displacement is $(2n)^{1/2} = 190$ steps. From Berg 1981.

a walk of over 18,000 steps. Total distance traveled (m) can be calculated as velocity time, thus m/s s. This relationship can also be expressed as:

$$\text{mean displacement} = (2 \times \text{velocity} \times \text{time})^{1/2}$$

This latter equation illustrates that the particle will move away as the square root of velocity or time, if we hold either velocity or time constant.

The key implication for stratigraphic studies is that where facies and patch migration is random in two dimensions, it will probably take a very long time for deposition to move out of an area and the local community,

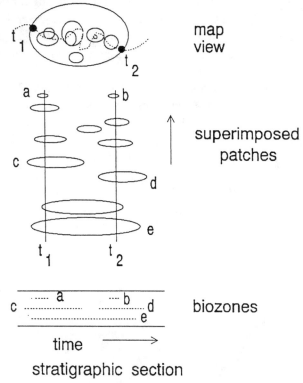

Figure 5.3 A biological community, such as a marine benthic community, consists of patches of different species (metapopulations) that are superimposed. Migration of the particle (patch migration over the core or section) therefore produces simultaneous preservation of many species, in the community, and in the section. Smaller patches tend to have shorter vertical ranges, as they pass over the section faster.

defined as an assemblage of coexisting, superimposed clusters of population patches (figure 5.3). The random walk of the particle will thoroughly scour the local area and its patches so that the same species will appear repeatedly in the section. We can conclude that the particle has finally moved out of the area when species with the smallest, most localized area occupied by their patches begin to disappear.

Testing the Particle-Patch Model

A stratigraphic section (the "particle") usually records population patches for only a fraction of each patch's duration, as it moves across the area of deposition. How can we quantitatively test, when sampling upsection, if the first several species to disappear represent patches that were the most locally distributed (figure 5.3)? One way is to use the observation that, within many taxa, the geographic range of a species (metapopulation) generally correlates with density of abundance (reviews in Ricklefs 1990; Hengeveld 1992). The most widely cited reason is that widespread species exploit abundant resources or ecological conditions (McNaughton and Wolf 1970). This correlation of areal range and abundance has also been found in fossils (see Koch 1991 for a review), where it is probably enhanced by sampling error of rare species.

Application of the particle-patch model to fossils allows us to use the range-density correlation as follows:

1. a stratigraphic unit is sampled near the bottom of a stratigraphic section or core;
2. the sample is completely analyzed for abundance data of component fossils, such as all benthic foraminiferal species;
3. similar samples are taken at intervals upsection; and
4. the abundance data from the upper samples are sequentially compared to that of the lowermost sample.

We would expect samples taken close to the lowermost (earliest) sample to be more similar in relative species abundance to that basal sample than would samples taken higher up because the population patches will have migrated farther away. Furthermore, if abundance is a proxy for area, the first species to disappear in a stratigraphic section will be the less abundant ones because they have the smallest geographic range.

Abundance Proportions and Isometric Plots

Figure 5.4 shows one way of quantitatively expressing the expectation that localized rarer species will be the first to migrate away from the area of deposition. The predicted proportion of each species (x-axis) is based on the proportion of it in the lowermost sample; the observed proportion of that species (y-axis) is what is actually recorded in each sample higher up. For example, Species A may comprise 10% of the total abundance of the lowermost sample (= predicted) and 5% of the total abundance of the adjacent sample above it (= observed). The line of isometry (slope = 1) indicates where the points would fall if the later sample has exactly the same abundance composition as the lowermost sample, i.e., each species is found in exactly the same proportion. We would expect sampling error around this line even for cases in which the true proportions were exactly the same, so the points would tend to fall on either side of the line. This effect can be corrected with binomial error bars $(npq)^{1/2}$ where n = total abundance, p = predicted proportion, and $q = 1 - p$.

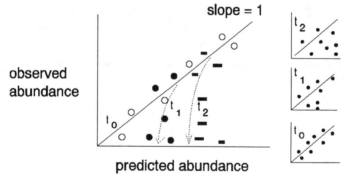

Figure 5.4   Each point represents the observed proportion of a species in a stratigraphic sample compared to its original proportion in the lowermost sample in the core or section. In samples higher in the section, rarer species are expected to deviate earliest from the predicted values, as in those which during the first time step (t_1). Because these rarer species generally have more localized patches, patch migration will take them away from the section first. Later periods will see larger patches, with higher density (proportions), migrate away (t_2). Because some rare species are widespread (Rabinowitz, Cairns, and Dillon 1986), they will persist in the section, as shown.

Figure 5.4 shows that, under the particle-patch model, we would expect species of low abundance (proportions) to be the first to drop away from the isometry line. This expectation is grounded in the abundance–geographic range correlation noted above. As the particle migrates away from its initial position where it produced the bottom sample of species assemblage, the first species to be "left behind" are those with the smallest geographic ranges, as indicated by the lowest abundance. As more time passes, recorded in samples even higher in the section, we would expect species of greater abundance, and hence geographic range, to also begin to diminish in proportion as the particle migrated even farther. The use of binomial error bars will help assure that observed diminishments are not solely attributable to sampling error wherein the less abundant species are merely unsampled. However, some observed points in figure 5.4 are purposely shown to not decrease to zero with time (upsection). These points occur because the correlation of abundance and area is only a general one; like many correlations, there are exceptions. Indeed, the widely cited work of Rabinowitz, Cairns, and Dillon (1986) on the meaning of "rarity" reflects this: some widespread species are rare and some very abundant species are highly endemic. Such species represent important deviations from the general correlation of abundance and area and must be accounted for in testing the model. Thus, some rare species may not disappear upsection, but keep reappearing, if they are widespread.

Another source of "noise" is that patches within any metapopulation are not of uniform abundance density, and vary in size and distance. Many, if not most, metapopulations exist as patch clusters, with a few large source patches surrounded by many small sink patches (Pulliam 1988; Murphy, Freas, and Weiss 1990, figure 3). More widespread species tend to have more and often larger clusters of patches. Each cluster of patches would, in turn, contain more and often larger patches, with all patches tending to be more densely populated. This implication of self-similarity will be explored later; the point here is its role in producing "noise" in the abundance-area correlation.

A second prediction that follows from the abundance–geographic range correlation is that species of lower average abundance would tend to have shorter vertical ranges: the smaller patches, or clusters of patches, would translate into shorter temporal durations as the particle moves across them (see figure 5.3). Again, confidence intervals are needed to insure that the shorter durations are not simply the result of higher sampling error in rarer species (Koch 1991).

A Test Using Benthic Foraminifera

An initial test of the particle-patch model used Paleogene benthic foraminifera (McKinney and Frederick, submitted). These organisms are well suited for such a test because they are diverse, abundant, and generally well preserved. The patchy distribution of living benthic foraminifera (Murray 1991) provides important modern analogs for understanding fossil patterns. The thick Paleogene strata of the U.S. Gulf Coastal Plain often provide an excellent record of many facies shifts owing to sea level change. Patch migration might therefore be expected as the various substrate-sensitive populations of benthic foraminifera respond to local water depth and other microenvironmental changes. The stratigraphic section studied by McKinney and Frederick records changes mainly in shelf deposits of intermediate depths. This was chosen as a transitional area that records a number of oscillating facies shifts, as opposed to deep sea sediments which are often deposited in more stable conditions.

Sequential stratigraphic samples were taken from a core drilled in south Georgia. (See McKinney and Frederick, submitted, for the full description and for more details and justification of the conclusions provided here.) The core consists of several hundred feet of carbonates from the uppermost Eocene and lowermost Oligocene in the eastern Gulf Coastal Plain. The core was drilled near the center of a Paleogene paleotopographic low called the Suwannee Strait (also Gulf Trough, and other names [Carter and McKinney 1992]). The Eocene-Oligocene boundary appears at about 976 feet down core. This locale was chosen for preliminary study because it was an area of active deposition and was a transitional area between the Florida Platform and more coastally influenced areas to the north (Carter and McKinney 1992). As an area of deposition, it can therefore be expected to record much evidence of facies and patch migration and environmental change. An interval across the Eocene-Oligocene boundary was chosen, too, because it is a time of special interest, and intervals on both sides of the boundary could be compared. Samples of the core were taken at approximate 3–5 foot intervals where possible, depending on lithology. After preparation and sieving, volumetric abundance counts were made of 102 species of foraminifera.

The results, shown in figure 5.5, generally conform to the particle-patch predictions of figure 5.4. Beginning at a core depth of 1,021 feet and ending at 915 feet, the predicted pattern emerges as the consecutive samples are compared to the original "base" sample at a depth of 1,032 feet (not shown). The sample (1,021 ft) closest to the base sample has observed abundances that deviate the least from those of (predicted by) the base sample. Going upsection, the correlation coefficient decreases as observed proportions deviate farther from those of the base sample.

Of great importance is the pattern of these deviations. As predicted by the particle-patch model, the single most abundant species (more than 100 individuals) in the base sample remains present and relatively common upsection, but finally begins to diminish in abundance toward the top. Conversely, the rarest species (fewer than 10) in the base sample show a rapid change in proportion in the adjacent sample (1,021 feet); 61% (19 of 31) of the rarest species disappear (figure 5.5 does not plot species with zero observed abundance, but percentages are shown). Interestingly, samples higher up the core show a decrease in this percentage, thus indicating that fewer of the rare species in the base sample are absent.

This seeming anomaly from the prediction does not, however, disprove the particle-patch model for two reasons. First, rare species are subject to much variation from sampling error (Koch 1991), which will significantly affect the calculated percentages. But this alone is a dubious explanation for the lack of a steady decrease of low abundance (and hence smaller patch) species upsection. The second reason may be seen in the ratios for each sample.

1,021: 19/19
1,017: 11/12
  939: 13/16
  915: 11/15

The denominator is the total number of base species that are absent; the numerator is the subset of these absent species that are the same as the ones that are absent in sample 1,021. Thus, most of the absent species are shared with the earliest sample, and indeed by all samples. This implies that rare, endemic species disappear early in the section, and generally stay absent, in accordance with the model. The other rare species, those that persist upsection, comprise very nearly half of each sample and would appear to be examples of species that are rare but widespread.

depth = 915

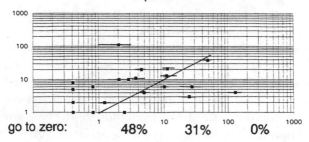

go to zero: 1   **48%** 10 **31%** 100 **0%** 1000

depth = 939

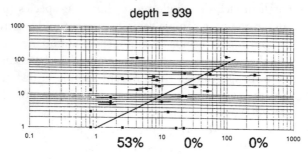

53% **0%** **0%**

depth = 1017

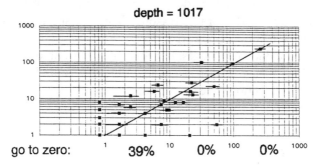

go to zero: 1 **39%** 10 **0%** 100 **0%** 1000

depth = 1021

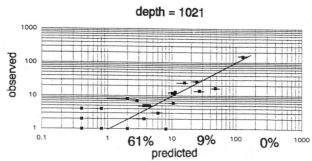

61% **9%** **0%**

predicted

Figure 5.5   Predicted versus observed abundance of foraminiferal species up-section in a core. Depth = core depth of sample, with the base sample at a depth of 1,032 feet. Solid line has slope of 1, indicating where predicted abundance = observed. 68% error bars shown. "Go to zero" percentages are the number of observed species whose abundances are zero, i.e., are absent from the sample but were present in the original base sample. The percentages are grouped in three abundance categories: 1–10, 10–100, and 100, reflecting their original abundance in the base sample. Predicted abundance was calculated by taking the total number of individuals in a sample and multiplying it by the proportional abundance of each species in the base sample. Thus, if species x composed 1% of the total abundance in the base sample, its predicted abundance in a later sample with 1,000 total individuals would be $0.01 \times 1,000 = 10$. Binomial error bars around this estimate (n p) were then calculated. "r" = correlation coefficient of regression.

Other features of figure 5.5 also conform to the particle-patch model. Species of intermediate abundance (between 10 and 100 individuals recorded) tend to deviate from the predicted at later samples upsection. But the percentage of new species does increase upsection, as predicted (figure 5.5). Most of these newly appearing species are rare, and thus ostensibly localized—also as predicted. A graphical agreement with the model is evident in figure 5.6, which shows the correlation coefficients for other samples upsection in the core, in addition to those samples in figure 5.5. There is a regular decrease as the "particle" migrates away from the base sample, perhaps in a random walk. But the Eocene-Oligocene boundary (at about 976 feet) represents a drastic change in species abundance and composition, caused by relatively rapid facies migration (deterministic, rapid particle movement) at this time (Prothero and Berggren 1992).

Sample 939 (figure 5.5), however, represents a return to conditions more similar to pre-Oligocene time, analogized by the particle returning to patches in closer proximity to earlier positions. After this, the coefficient steadily decays, indicating a return to a random particle walk, away from the base sample position. Another agreement with the model is that species in the core show a highly significant correlation between mean abundance in the samples and duration, defined as consecutive samples where the species in question is present. This indicates that more abundant species tend to have larger patches or clusters of patches because, under the particle-patch model, such patches or clusters will take more

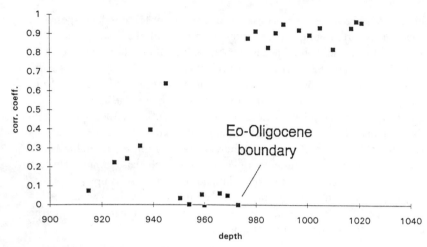

Figure 5.6 Correlation coefficients of predicted versus observed abundance plots (e.g., figure 5.5 as shown), as a function of core depth. Coefficients nearer to 1 indicate nearness to isometry, i.e., overall species proportional abundance is similar to that of the base sample (depth = 1,032 feet).

time to migrate over the area of deposition, producing an extended vertical range (duration).

Implications of Population Patches for Paleobiology

Viewing a stratigraphic section as the record of patches migrating over the area of deposition has many implications. We now discuss two sets of implications: (1) taphonomy, biostratigraphy, and paleoecology, and (2) fine-scale study of speciation and extinction dynamics. The first set of implications are only briefly discussed here. (McKinney and Frederick, in a submitted paper, explore these further).

Implications for Taphonomy, Biostratigraphy, and Paleoecology

A basic implication of population patches for taphonomy is that spatial variation is conflated with temporal variation in abundance, morphology, and other population and community data. This is quantified in what we have termed the *index of patch resolution* (IPR), which is the organism preservational rate (dP/dt) divided by the particle (patch) velocity (dH/dt). If we include deposition of sediment and other fossils that comprise the matrix (dS/dt), then

$$\frac{\left(\dfrac{dP/dt}{dS/dt}\right)}{(dH/dt)} = \frac{\text{volumetric abundance}}{\text{particle velocity}} \qquad \text{(eq. 3)}$$

Two more implications of population patches for taphonomy are the effects of: (1) variable rates for all three variables (P,S,H) and (2) postmortem processes that distort the original (biocoenotic) abundance information in the taphocoenosis. Both effects are summarized in figure 5.7. Postmortem effects are shown to reflect abundance as follows: Original abundance (A) is the number of individuals that inhabit an area per unit time. A subset of these (M) may die per unit time and a subset of those that die may be deposited (D). Finally, the subset of those deposited that are preserved (i.e., not recrystallized or otherwise destroyed) is P.

Much of the theory and data presented on the particle-patch model is based on the assumption that P provides at least some indication of A; patchiness is most evident if large dense patches are recorded as

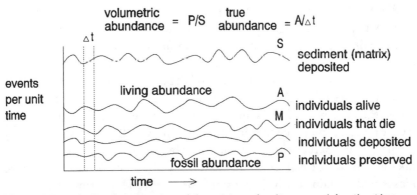

Figure 5.7   Idealized time series of deposition of sediment and fossils. Abundance of preserved fossils (P) is a subset of the original life abundance per unit time (A); this idealized relationship depicts P to be roughly proportional to the original abundance. Because deposition of fossils (D) and sediment (S) is known to be episodic with many nondepositional periods (e.g., Sadler and Strauss 1990), dt is assumed to be long enough to produce a nonzero value.

such, compared to small sparse patches. More rigorously, it assumes that $M = kA + \text{error}$, $D = kM + \text{error}$, $P = kD + \text{error}$ (where k is a proportionality constant and error is taphonomic "noise"). Summing the equations by insertion yields $P = kA + \text{error}$. Because this error is the sum of all the errors (stochasticity) found in each transition between A and P, it could obviously be quite large. This method also ignores such potential problems as vertical mixing and transportation and deposition of nonlocal fossils.

There are, however, four reasons to suspect that the errors may not always completely destroy the original abundance signal. First are the empirical results we have already discussed. It is difficult to explain the conformity of data with model predictions if fossil abundance is, say, random with respect to original abundance. This is because the model is based on the assumption that area is indeed correlated with abundance. Second, various mathematical trials indicate that comparisons of P among species are robust to high error, provided the original abundances among the species were significantly different (McKinney and Frederick, submitted). Thus, a species that is 100 or even just 10 times more common will very likely be significantly more abundant as a fossil even when preservational rate of all species has much random variation around some proportional mean.

A third reason to suspect that the presence of error need not always destroy the abundance signal turns on qualitative and theoretical arguments. For instance, we know that the value of k in all the equations is likely to be taxon-specific; species will differ in mineralogy, habitat, and other ways that affect rates of death and deposition and preservation. But these traits are relatively invariant within a species, so that relative abundance through time of the same species will have relatively constant intrinsic biases. A major change in abundance may well at least partly reflect a true nonartifactual change. Similarly, abundance comparisons among taxa may be distorted by differing taxon-specific traits, but these intrinsic biases will remain relatively constant through time. Extrinsic biases are minimized because all species in each sample were deposited in the same environment. None of these points denies that mixing, time averaging, size sorting, abrasion, and other well-documented distortions occur (Allison and Briggs 1991; Kidwell and Behrensmeyer 1993). Many of these biases, however, act upon all fossils within a given vertical unit and, again, distort the abundance signal in a consistent way.

Fourth, and perhaps most encouraging, are estimates that within-habitat time averaging for shelves is on the order of decades to perhaps 10,000 years (Martin 1993; Kidwell 1993; Kidwell and Bosence 1991). In environments such as that studied by McKinney and Frederick, which was a massive carbonate shelf, long periods of uniform conditions may result in time averaging of a few thousand years and perhaps much less. Time averaging may not hinder useful study and interpretations if migration of microenvironments and patches (dH) is small, especially given a random direction of migration, which often takes an enormous amount of time to move very far in one direction (figure 5.2). During thousands of years, the particle will pass through many patches (including those of the same species), which may become time averaged; but rare species with smaller patches should be distinguishable from common, widespread species. While single patches will not be detectable under such conditions, relative patchiness among species should be. Even vertical mixing—mixing over a few thousand years of deposition—would not necessarily destroy a long-term signal; e.g., rare and patchy species would likely appear as rare and patchy throughout the section. The open question is the degree of patch resolution visible, and this is largely an empirical issue: on what scales (temporal and spatial) is a species patchy, and how fast was deposition relative to this scale? We are hopeful that ongoing taphonomic work, such as that on benthic foraminifera discussed by Martin (1993), will incorporate the role of patchiness and patch migration in producing taphocoenoses.

Biostratigraphic implications of patch deposition are also numerous, especially given that pelagic and planktonic organisms are patchily distributed in the water column. Biozones and many other biostratigraphic time series do not consist of a single population of a species. Rather, a biostratigraphic sequence is the product of many patches, of varying connectedness, size, and abundance density and uniformity. The particle-patch model provides an analytical tool for determining direction and rate of facies and patch migration in the geologic record. The greater the accuracy of dating available for biozone and stratigraphic boundaries, the greater is the accuracy (and the finer the scale) at which facies and patch migration can be determined.

Classical biostratigraphic correlation is the correlation of different "particles"; widely distributed patches of the same species can be used to correlate particles that were traveling through those patches at the same time. Each stratigraphic section is a set of superimposed patches, with only a very short time interval of each patch's existence being recorded. Even if we examine many stratigraphic sections, using fine-scale temporal correlation and analogized by many particle walks through patches of the same species, only a tiny fraction of even large patches will be seen, and these for only a fraction of their original duration. Rare species with small and rapidly migrating patches are the most likely to present correlation problems such as determining true extinction boundaries, including such well-known sampling effects as the Lazarus and Signor-Lipps Effects (Marshall 1991).

Patch deposition models have considerable implications for studies of paleocommunities. Fossil species are rare in the same log-series pattern seen in living species, and rarer species are more locally distributed (Koch 1991). This pattern has been applied to a number of important practical problems, such as calculation of sampling error when comparing different strata or paleocommunities on the basis of species composition (Koch 1991; Carter and McKinney 1992). The particle-patch model supplements such work by attempting to describe, in detail, how the abundance patterns seen in the record are actually produced.

Communities are sets of superimposed coexisting patches of different species. In figure 5.3 we may ask whether the particle is in the same set of patches in both communities, with differences due only to sampling error of rare species, or whether migration of the smaller patches with rare species out of the depositional area has caused a true shift in the species that compose the community. Miller's (1988) work on subfossil deposition is a precursor to this view in showing how lateral abundance variation in a community may be preserved. Koch (1987, 1991) has emphasized that rare species are found at fewer geographic sampling locales mainly from sampling error, but the particle-patch view illustrates how this error may only exaggerate the abundance-area correlation where rarer species are often more locally distributed.

The particle-patch model is also useful in describing the paleocommunity abundance patterns calculated by Fursich and Aberhan (1990), who discuss how rank abundance of the more abundant species is relatively stable through time. Figure 5.5 shows that the single most abundant species stays most abundant for a very long time because it is also the most widespread. A species may thus remain abundant despite some facies and environmental shifts. Similarly, the numbers of individuals recorded for species of intermediate abundance does not change much until relatively high in the section. Species of high and intermediate abundance will tend to have stable abundance patterns even if environmental shifts occur.

Implications for Fine-Scale Speciation and Extinction Studies

By fine-scale studies we refer to data on patch dynamics derived directly from local stratigraphic sections or cores. Any complete study of speciation, or extinction, must address population change as the basic ecological and phylogenetic unit (Eldredge 1992). But "population" is a relatively meaningless term except in the context of gene flow and spatiotemporal morphological variation. Indeed, perhaps the main message of metapopulation theory is that a "population" occupies a continuum of entities ranging from a completely isolated collection of individuals to near-total genetic exchange with other entities. Previous fossil studies of speciation have focused only on temporal morphological variation, and where the term "population" is used at all, it is in a vague, often erroneous way. The value of incorporating patchiness and metapopulations is in providing a precise description and theory of spatial variation of abundance, which can be extended to include spatial morphological variation by statistical analysis of individuals and their traits.

On the negative side, the taphonomic problems we have already noted, combined with the problem of patch migration, will obviously limit direct observation of patch dynamics to well below that seen by a neontologist in the field. Many key aspects of metapopulation dynamics, such as patch equilibria (Whitlock 1992) or correlated patch extinctions (Harrison and Quinn 1989) on yearly time scales will likely be hopelessly beyond study in the record. For example, population dispersal and growth to carrying capacity both largely occur at rates on the order of decades or centuries that render the fine details invisible (figure 5.8).

Spatial dispersal of species can be modeled as a diffusion process; such models have proven to be relatively accurate when tested with empirical data on introduced species (Murray 1988). One of the most accurate is the advancing wave model: $V = 2(rD)^{1/2}$, where V is the speed of advance, r is the net birth rate (from the familiar logistic intrinsic rate of increase), and D (square meters per yr) is the diffusion coefficient. Individual organisms are considered as particles, so D can be most usefully expressed

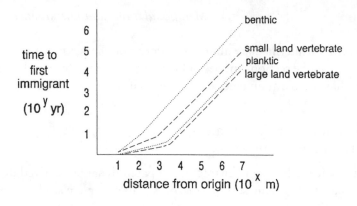

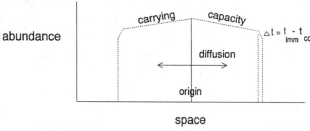

Figure 5.8 *Bottom:* Calculations from the "advancing wave model" demonstrate that organisms generally reach local carrying capacity abundance (t_{cc}) soon after the first immigrants of the wave arrive in an area (t_{imm}), compared to the time it took to reach the area. Therefore, the diffusion process may best be viewed as an expanding "shell" with high walls. Carrying capacity is shown to decrease away from the origin, in agreement with empirical evidence that conditions tend to become less optimum toward the edges of a species range and away from where it evolved. *Top:* Order of magnitude estimates of how long various types of organisms take to disperse in a circle with radius as distance (meters) from origin. Benthic estimates assume no planktonic dispersal stage. Benthic estimates based on individual velocity (I) of 1 m/day, step size (S) of 0.01 m, net birth rate (r) of 0.05. Small land vertebrate: I = 100 m/day, S = 1 m, r = 0.0007. Planktonic: I = 100 m/day, S = 0.1 m/day, r = 0.1. Large land vertebrate: I = 1000 m/day, S = 10 m, r = 0.0005. Values of r were taken from standard graphs that estimate it from fecundity, body size, and other correlates of birth rate; I and S are scaled to reflect known organisms in each category, but will obviously vary. The general pattern of the graph is relatively robust to realistic changes in all these parameters. The basic inference is that, barring strictly benthic forms, dispersal can occur with surprising speed (a few 100 to 1,000 yr) over long distances.

as individual velocity, I (e.g., m per day) multiplied by S, step size (the mean distance an individual will move in one direction before changing direction). This model thus incorporates rate of movement plus rate of reproduction. It also incorporates patchiness; as Murray (1988) discusses, patchiness can arise from the destabilizing effects of diffusion gradients.

Similarly, estimates from population genetics show how morphological variation in patches will generally appear "instantaneous." Small populations tend to have enhanced morphological variation from decreased genetic variation and increased homozygosity. When enhanced morphologic variation is apparent in the fossil record, it might thus be taken as evidence of a new, isolated patch being established. However, population geneticists have shown that a small population will recover most its genetic variation by spontaneous mutation quite rapidly in terms of geologic time (Lande and Barrowclough 1987). For example, quantitative, polygenic traits will generally be recovered in just a hundred to a thousand generations (table 5.1). Given a generation time of one to a few years for most marine invertebrates, recovery will occur in the range of a hundred to a few thousand years. Most evolution occurs from such polygenic changes (Lande 1981). But where evolution is influenced by single genes of large effect, we may see some post-isolation effect in the record because such genes have a much lower mutation rate. As a result, it may take more than a million generations to recover full variation in such genes.

The relatively rapid recovery of genetic variation in small isolated populations is enhanced by the surprising difficulty with which genetic variation is removed. Even if only a single mating pair begins the population patch, only about 25% (at most) of the genetic variance will be lost

Table 5.1
Population Size and the Maintenance of Genetic Variation

Type of Variation	Nature of Selection	Population Size	Recovery Time (generations)
Quantitative	Neutral	500	100–1000
Quantitative	Stabilizing, or fluctuating optimum	500	100–1000
Single-locus	Neutral	10^5–10^6	10^5–10^7

SOURCE: Modified from Lande and Barrowclough 1987.
  NOTE: Population sizes needed for mutation to maintain significant amounts of genetic variation at equilibrium.

(Lande and Barrowclough 1987). Larger founding populations will of course lose even less. Instead, it takes protracted stress to produce sustained small populations (and thus inbreeding) or repeated population crashes to reduce genetic variation more severely.

But on the positive side, coarser aspects of metapopulations and patches should leave considerable fossil evidence. As an extreme case, we would hardly expect a rare, very patchy species to leave the same kind of fossil record as a very abundant species with a more homogeneous distribution. Yet instead of designing speciation studies to collect such data, paleontologists have generally ignored patchiness. What do detailed morphometric studies, such as that of Williamson (1987), Sheldon (1987) or many others, really mean when we consider that the stratigraphic sections analyzed are the time-averaged product of migrating patches? Even in the case of planktonic organisms, patchiness should be considered. That we are often observing migrating population patches could help explain Lande's (1986) observation that nearly all detailed fossil sequences have substantial fluctuations in average morphology at the smallest resolvable time scales.

There are many ways that paleobiologists can treat the influence of patches on fossil deposition as crucial information, instead of as a depositional artifact that distorts the record. Such information should be built into theoretical expectations. For example, Sheldon (1993) suggested that species adapted to unpredictable environments, such as shallow waters, may show more punctuated patterns of morphological change than species in stabler environments (figure 5.9). But tests of Sheldon's hypothesis must incorporate the effects of patches and their migration. In comparing a stratigraphic sequence to these patterns, we must consider that at least some of the vertical patterns of variation in both abundance and morphology may be produced by patch migrations over the area of deposition. Does a rapid stratigraphic change in morphology simply represent deposition of a different cluster of patches in the metapopulation? Unfortunately, the more patchy metapopulations—those most likely to show this artifact—are also the ones most likely (according to traditional allopatric models) to rapidly speciate.

Metapopulation theory also contributes to one of Sheldon's (1993) key predictions—that disturbance-adapted species with short-lived populations may have enhanced species longevity. Evidence that shallow water species are more widespread and eurytopic (Jackson 1974) implies that

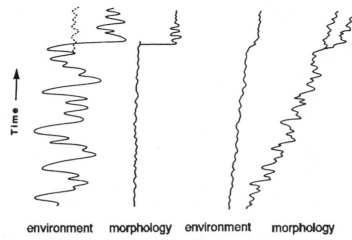

environment   morphology   environment   morphology

Figure 5.9   Sheldon (1993) has suggested that species adapted to more frequently disturbed environments (*left*), such as shallow waters, will be more broadly tolerant and show little morphological change, compared to species adapted to stabler environments (*right*), such as deeper waters. Modified from Sheldon 1993.

their short-lived populations may indeed leave a more punctuated fossil record, but with greater species longevity. There is some support for this inference in one of Levin's original metapopulation equations used to predict species duration from population duration:

$$\mathrm{Tm} = (\mathrm{Tl})^{\exp(Q^2/2[M-Q])}$$

Here, Tm = species duration, Tl = mean population duration, M = number of habitat patches, and Q is the equilibrium number of habitat patches. While this equation is simplistic, in assuming—among other things—equal size patches (Hanski 1989), one general point is almost certainly valid: species duration observed in the rock record increases only arithmetically with population duration but exponentially with the number of patches occupied. Thus, more widespread disturbance-adapted ("opportunistic") species could persist in the record much longer than species with fewer but longer-lasting patch populations. One could test this in the fossil record by comparing Tm, as recorded in the literature, to the local stratigraphic duration of local populations as an estimate of Tl. We see here, however, that Tl is only the time it took the particle to migrate across the population patch (i.e., the time it took the patch to migrate over the area of deposition of the stratigraphic section).

Fossil patchiness can also be used to directly test theories of speciation dynamics. For example, it has often been noted that patchy, rare, or poorly dispersing species should have higher rates of speciation and extinction (Endler 1989; Stanley 1990; Eldredge 1992). This correlation could be quantitatively tested by statistically characterizing abundance variation and mean abundance in the stratigraphic record for each species. Metrics of abundance variation could include standard deviation, coefficient of variation, and serial correlation of the taphonomic series to measure volatility. Figure 5.10 shows how one might use this method to discern species that have very heterogeneous abundances ("patchy" in time and space). Then, to see if there is a correlation between patchiness and rates of speciation and extinction, abundance variation would be compared to data from the published literature on geographic range, the total known geologic range of each species, and perhaps number of closely related species.

As a final example, fossil patch data might be used to test the theory that moderate environmental stress tends to maximize speciation rates, in part by maintaining genetic and phenotypic variation (Parsons 1991, 1993). It is well known that stress increases toward the boundaries of species geographic ranges. As the outer areas of a metapopulation migrate over the area of deposition, we might therefore expect to see increased morphological variation and increased patchiness (smaller, sink patches tend to occur toward the limits of a species' environmental tolerances [Pulliam 1988]). In a stratigraphic sequence, this expectation could translate into more abundance and morphological variation at the beginning and end of the species range in the sequence as the metapopulation moves across the area.

This increasing amplitude of the vertical abundance series might be seen as a "red shift," a term used by Steele (1986) and Pimm (1991) to refer to increasing variation (wave amplitude) in a time series (figure 5.10). Red shifts would be most common in metapopulations with relatively small geographic ranges because their species boundaries would transit local depositional areas more often than would the boundaries of widespread species. McKinney and Frederick (1992) found tentative evidence that red shifts in abundance were indeed more common in fossil

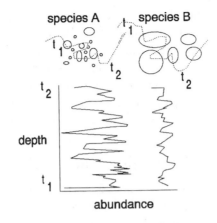

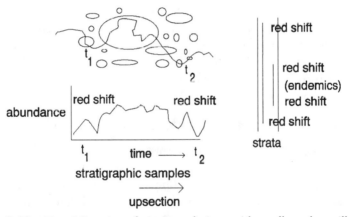

Figure 5.10 *Top:* Migration of metapopulations with small patches will produce a stratigraphic abundance pattern that is more highly fluctuating than a metapopulation with a less patchy distribution—that is, with larger, more uniformly dense patches. *Bottom:* Stratigraphic abundance may often become more variable ("red shift") toward the beginning and end of a species' stratigraphic range if the stratigraphic section records the migration of the smaller, more fragmented satellite ("sink") patches that surround the larger core ("source") patches of most metapopulations. More localized, endemic metapopulations may be more likely to record red shifts, as red shifts are recorded when the fringes of metapopulation patches migrate over a local area of deposition.

foraminifera species with shorter durations in the stratigraphic sequence, as would be expected if the forams had smaller geographic distributions. The red shifts were interpreted using Pimm's (1991) findings that increasing abundance variation in a population leads to early extinction of the population. Because patch migration from environmental shifts is inevitable over the long time spans observed in a stratigraphic sequence, this process is a more plausible cause of the apparent red shift.

Disturbance Dynamics: From Patches to Biodiversity

The second part of this paper outlines a model, with evidence and implications, that relates patch disturbance to origination and extinction of species and higher taxa. Despite the name, the model recognizes that most change is derived from the complex interplay of biotic and abiotic factors. It thus does not rely solely on abiotic or extrinsic factors to drive change.

Disturbance, as defined here, refers to any abiotic or biotic change (such as exotic species invasions) that is of an unusually high magnitude. Intrinsic factors are incorporated into this view by noting that intrinsic (taxon-specific) traits will determine whether a group will be affected by a given disturbance.

Intermediate Disturbance Model of Maximal Speciation

Allmon (1992) discussed a model, shown in figure 5.11, in which intermediate levels of disturbance can maximize speciation rates by fragmenting metapopulations into isolated populations. The disturbance further enhances speciation by its effect on persistence and differentiation of the isolated populations, according to Allmon's (1992) three-stage process of allopatric speciation: isolation, persistence, and differentiation. Lesser degrees of disturbance produce little change, whereas greater degrees cause extinction. In terms of metapopulation theory, the increasing disturbance vector represents increasing correlation of patch extinction, as explored by Harrison and Quinn (1989). This correlation is distinct from the well-known intermediate disturbance hypothesis of diversity maintenance (Petraitis, Lathan, and Niesenbaum 1989), which posits that maximum (intra-community) diversity occurs at intermediate disturbance levels. The reasoning underlying the latter hypothesis is that too little disturbance permits dominance by a few competitively and otherwise interactively

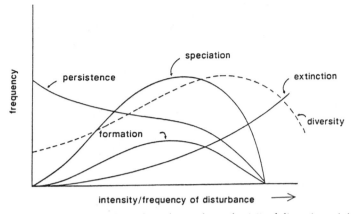

Figure 5.11 An "intermediate disturbance hypothesis" of diversity origination. Rate of origination is highest at intermediate levels. This is where rates of isolate formation and isolate persistence are highest. Modified from Allmon 1992.

superior species, whereas too much disturbance reduces diversity from high environmental stress.

In contrast, the model presented here addresses diversity origination. We designate this new model as the *intermediate disturbance model of maximal speciation*. It is an elaboration of Stanley's (1986) "fission effect." The model extends his and Allmon's (1992) ideas in many ways we shall discuss, but three key extensions are: (1) it focuses on biodiversity creation and loss over evolutionary time scales, rather than maintenance over ecological scales; (2) it expands the "fission effect," which emphasized biotic causes of population fragmentation, to include all types of disturbances; (3) it is a scale-free model that is applicable to many spatial scales and that draws on the many "hollow curves" (fractalness) documented in many biotic and abiotic patterns.

Figure 5.12 (bottom) illustrates the distinction between disturbance extent and intensity. *Extent* refers to geographic extent (area) of a disturbance; *intensity* refers to the degree of disturbance in the area affected (deviation from norm in one or more environmental parameters). For example, a minor salinity change has less intensity than a major one, whatever the extent. Lower intensity disturbances will generally result in lower rates of peak speciation and will induce speciation in fewer taxa, whatever the extent. This is because isolate formation, persistence, and differentiation will occur only in those taxa that are sensitive to the dis-

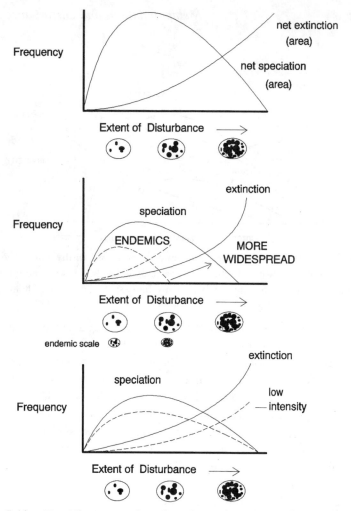

Figure 5.12 *Top:* The intermediate disturbance hypothesis of maximal speciation states that, as areal extent of disturbance increases, there will be a peak of
speciation (within the disturbed area) at intermediate levels of disturbance.
Middle: Endemic species (broken lines) will tend to speciate more often than
will widespread species (solid lines) because (more frequent) disturbances of
smaller extent will cause more fragmentation of their total ranges. Endemics
will thus be subjected to (what they perceive as) intermediate levels of disturbance more often. *Bottom:* Whereas extent is the area affected ("breadth" of
disturbance), intensity is the "depth" of disturbance, i.e., the number of environmental parameters (salinity, temperature, etc.) affected in the area and the
magnitude that each is affected. Lower intensity disturbances (broken lines) will
tend to affect fewer species.

turbance (i.e., stress). A biotic metric of intensity is therefore the number of different taxa that are affected. Catastrophic intensities will affect nearly all organisms.

Figure 5.12 (middle) also illustrates what happens when intensity is held constant at a high level and disturbance extent varies from local to regional scales. The key feature is that this scenario can account for differences in geographic range among species: narrowly distributed species will tend to have their highest speciation rates from relatively local disturbances compared to widespread species. This occurs not only because of the greater deployment of the latter but because of their generally better dispersal. As extent of disturbance increases, an increasing number of widespread species will experience population isolation and differentiation. Isolation and differentiation owing to disturbances of great geographic extent peak at an intermediate disturbance intensity, causing extinction of even widespread species. In addition to varying position, the shape of the speciation and extinction curves will vary.

The net result of these interactions of extent and intensity is portrayed in the top graph of figure 5.12, which shows the speciation peak that occurs when disturbance effects on all species, with varying geographic ranges, are summed across a range of areas. At very high intensities and extents, such as the global end-Permian event, mass extinction occurs and speciation is greatly depressed during the disturbance (Erwin 1993). This kind of event qualifies as a "press" disturbance, which persists for a long time, rather than a "pulse" disturbance, which perturbs the system and is gone, although its effects may linger (Underwood 1989). The K-T extinction may qualify as the latter, if the bolide impact hypothesis is true.

Theoretical Evidence: Explaining Fossil Patterns at Many Scales

Many fossil biodiversity patterns, both phylogenetic and ecological, conform to the intermediate disturbance model. One of the most important examples is seen in the well-known paleontological tendency to explain biodiversity in terms of abiotic causes (Donovan 1989; Signor 1990; Raup 1992). Cracraft (1992) has discussed in detail why origination and extinction events that occur across large spatial scales are generally caused by physical phenomena, such as tectonics and climate. In terms of Harrison and Quinn's (1989) model of correlated events, we would expect a correlated disturbance that affected many taxa throughout a large area (i.e., high intensity and extent) to be physical in nature because biotic disturbances such as the predation in Stanley's (1986) original fission effect would tend to be more localized and selective (discussion in Cracraft 1992).

That the most intense and extensive biotic disturbances tend to affect isolated biotas illustrates the special conditions needed to produce correlated biotic disturbances. Vaporization by a large bolide would thus exemplify the ultimate correlated event in the vaporized area, representing Raup's (1991a) "field of bullets" selection in which traits are irrelevant. Correlated events of less intensity would promote Raup's "wanton selection" wherein some traits promote survival under the new conditions. This view also conforms to evidence cited by Brooks (1991) that most speciation events are from vicariance, that speciation rates are controlled by geological change, and that the larger the spatial scale, the older will be the events producing the patterns seen.

The disturbance model also agrees with what Stanley (1979) has called the "first law of macroevolution": the empirical correlation of taxonomic origination and extinction rates among clades seen in the fossil record (see also Stanley 1990). For the model to be valid, it must explain two key properties of this empirical correlation. First do originations exceed extinctions in all taxa, thus producing a regression line with y-intercept greater than 1 (figure 5.13). Such a slope would be explained by the fact that disturbance frequency decreases with magnitude (in a hollow curve fashion, as discussed shortly). The result is that speciation-causing intermediate levels of disturbance will always exceed in frequency of occurrence the more extreme extinction-causing levels of disturbance over long time spans. This process is scale-free in that longer time spans will, on average, produce larger disturbances, affecting larger geographic areas and more widespread metapopulations. Thus, figure 5.12 is (in theory) valid across all scales, in that speciation will always tend to exceed extinction: at any specified scale, smaller disturbances will (on average) be more frequent than larger ones so that intermediate-level speciation-causing events will outnumber more extreme extinction-causing ones. Endemic species can therefore be expected to speciate (and become extinct) more than will widespread species.

Such self-similar disturbance-driven dynamics, acting across many spatial and temporal scales, may be a prime driver of the large number of

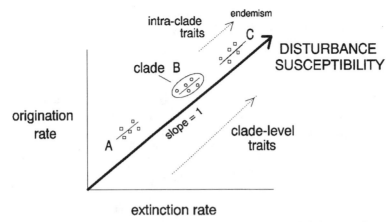

Figure 5.13   Graphic depiction of how origination rates slightly exceed extinction rates within and among clades. The vector that runs parallel to the line of isometry to the right represents increasing turnover. Intraclade traits such as endemism increase turnover; interclade traits such as disturbance susceptibility increase turnover in a more general ("structural") way, being based on more deeply rooted phylogenetic (e.g., class-, order-level) differences.

causes and consequences of metapopulation dynamics that have hollow curves in plots of frequency versus magnitude (table 5.2). In terms of intrinsic dynamics, Hanski and Gilpin (1991) have noted that metapopulation structure may be hierarchical, with a large population composed of subpopulations, which in turn are composed of patchily distributed individuals. Thus, when patches of a metapopulation are digitized, the patch size-frequency distribution is a hollow curve, no matter what the size of the metapopulation.

Similarly, Frontier (1986) examines the fractal nature of individual abundance in communities. Metapopulations (species) in a genus, along with genera in a higher clade, show the same hollow curve pattern (table 5.2). Regarding disturbance, Raup (1992) attributes the hollow curve of extinction rates to the hollow curve of bolide sizes; but a vast number of disturbances also have this frequency distribution (e.g., Dott 1983; Hsu 1989). Thus, curves that sum the coincident effects of different disturbances happening simultaneously would be hollow.

Hollow curves are very abundant in nature; they indicate 1/f, or fractal, patterns that result in power laws (e.g., in frequency versus magnitude plots) whose constants indicate self-similar patterns (Schroeder 1991).

Table 5.2

Power Law Distribution of Disturbance Effects on Metapopulations

Disturbance Dynamic or Effect Having Hollow Curve	References
Extrinsic dynamic:	
Disturbance	Dott 1983, Hsu 1989, Raup 1992
Intrinsic dynamic:	
Patch size (per metapopulation)	Whitlock 1992, Hengeveld and Haeck 1982
Patch area (per metapopulation)	Murphy et al. 1990, Erickson 1945
Temporal abundance variation (per population)	Pimm and Redfearn 1988, Pimm 1991
Species area (per clade)	McKinney (in preparation)
Genus area (per clade)	Flessa and Thomas 1985
Effect:	
Species longevity (per clade)	Levinton 1988
Genus longevity (per clade)	Raup 1991a
Extinction rate (biosphere)	Raup 1991b
Ecosystem cascades	Plotnick and McKinney 1993
Subtaxon per taxon	Burlando 1990
Phylogenetic branching	Harvey et al. 1993
Ecosystem diversity and per-taxon abundance	Magurran 1988

NOTE: The dynamics and results of disturbance acting on metapopulations often have hollow curves in frequency versus magnitude plots. These translate into power laws that imply self-similar dynamics across many spatial and temporal scales. References are illustrative, not exhaustive.

But such self-similarity is rarely indicative of a simple, single cause; rather, these distributions are often produced by many causes interacting in complex ways. However, whereas log-normal distributions are produced from multiplicative interactions, the hollow curves of 1/f patterns are produced by more synergistic interactions that amplify some events in "supermultiplicative" ways (West and Shlesinger 1990). Examples include higher taxa that are superrich in lower taxa and metapopulations that are superabundant or far more widespread than any others in their clade.

Because of an often complex origin, it is difficult to specify exactly what all this self-similar scaling means, except that it provides an intriguing common thread to relate processes at different spatiotemporal scales. This will be explored later in this paper, when we show that even self-similar patterns do not necessarily extrapolate simply to higher levels (McKinney and Drake, in press).

The second property of the empirical correlation between origination and extinction calls for an explanation of the location of points along the regression, i.e., why some clades have high turnover while others have low (figure 5.13). Stanley (1990) cited a number of clade properties that can affect turnover: behavioral complexity, stenotopy, dispersal, and population size. In terms of our intermediate disturbance model of maximal speciation, these properties in turn determine turnover by influencing two key higher-level properties: geographic distribution and how sensitive the clade is to environmental changes.

As noted, a less widely distributed clade will be more frequently disturbed because smaller disturbances are more common. Endler (1989) reviews how restricted distribution promotes both speciation and extinction (also see review of the many authors noting this in Flessa and Thomas 1985; Kunin and Gaston [1993] report a correlation between rarity and poor dispersal, implying a synergistic effect). Traits that increase sensitivity to environmental change include behavioral complexity and stenotopy. Clades composed of species with these traits are more likely to be disturbed because they are affected by smaller changes in more kinds of environmental parameters, such as salinity, temperature, and co-occurring biota. Eldredge (1992) discusses increased extinction and origination with stenotopy. Vrba (1987) reports mammal clades in which stenobiomic groups had higher speciation rates than did eurybiomic. Similarly, Baumiller (1993) shows that "specialized" pinnulate crinoids had higher turnover than did non-pinnulate.

Overall, highest turnover will be predicted in clades with the most localized/restricted distributions and with the greatest sensitivities to environmental change. The self-similar nature of disturbance implies that these predictions apply to all groups and for comparisons at any taxonomic level. For example, endemic taxa within a clade should show higher turnover (figure 5.13).

Empirical Evidence: Gastropods from the Gulf Coast Paleogene

So far we have considered evidence for our intermediate disturbance model's prediction that speciation and extinction should be correlated within and among clades. Yet the model also predicts a speciation-extinction correlation for regional or global events: an intermediate disturbance of a given area can cause extinction of endemics, while causing increased speciation among more widespread species (figure 5.13). The result is an increase in both extinction and origination when there is a disturbance that is of an intermediate magnitude *for the size of the area observed*. For example, Allmon et al. (1993) and Jackson et al. (1993) are among those reporting high extinction and origination during a regional extinction episode (see also Flessa et al. 1986). We explore further evidence for this correlation here. We also note that this type of study is an excellent illustration of how paleontologists can overcome the problem of patch migration in single stratigraphic sections in order to examine speciation and extinction at the regional scale of the metapopulation itself.

At regional scales, stratigraphic records are usually of low paleontological completeness (*sensu* Allmon 1989). Data for studying speciation in such low completeness stratigraphic situations therefore consist of the distribution in time and space of first appearance datums (FAD's) of the species within clades of interest. This contrasts with usual studies of closely spaced samples of varying morphology. An observed pattern of FAD's (figure 5.14) is some combination of two conditions: it can approximate the actual time and place of species origin, or it can be an artifact that records a later time and place than the origin. Given any pattern of FAD's, the two relevant questions are: (1) how accurately does the pattern reflect the true pattern of first appearance of species in this group? and (2) can any useful conclusions be made about the processes underlying this pattern?

Turritellid gastropods are a good macrofossil group in which to study speciation dynamics in a relatively incomplete section. They are common in Mesozoic and Cenozoic marine molluscan assemblages. They are taxonomically diverse and appear to evolve rapidly (e.g., Kauffman 1977) and are widely used biostratigraphically (e.g., Toulmin 1977; Saul 1983). They are diverse and common in Recent communities, allowing ecological comparisons with fossils.

In the example discussed in this section, we argue that:

1. Turritellid gastropods of the Paleogene U.S. Coastal Plain show a first occurrence pattern that suggests exceptionally high origination rates during marine regressions.
2. Peramorphic heterochrony was a main mode of evolution, and much of it may have had little adaptive significance.

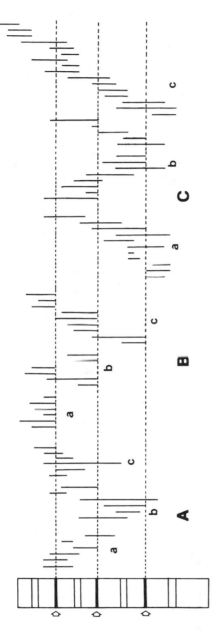

Figure 5.14 Three examples of patterns of first appearance datums and possible interpretations. Vertical lines are observed temporal ranges of species. In the simple hypothetical stratigraphic record shown schematically on the far left, different rock types are indicated; major hiatuses marked by unconformities are shown by thick black lines and arrows. (A) Species in three clades (a, b, c) show no pattern of differential first appearance with respect to either lithologic unit or unconformities, making causal interpretation of extrinsic factors difficult. (B) Almost all species in three clades show first appearances immediately following hiatuses. An unknown proportion of observed first appearances are artifacts of the incomplete record, and biological differences among the clades cannot be inferred. (C) Species in three clades differ in their pattern of first appearance. Species in clade "a" show most first appearances in the thicker units (perhaps limestones); species of clade "b" show most first appearances in the thinner units (perhaps shales); species in clade "c" show most first appearances immediately following major hiatuses, suggesting that many or most of its species actually originated during this interval.

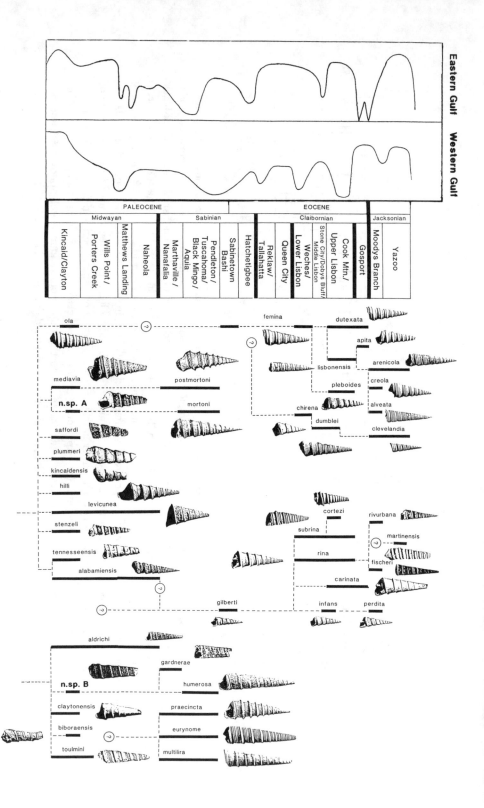

Eastern Gulf

Western Gulf

PALEOCENE		EOCENE	
Midwayan	Sabinian	Claibornian	Jacksonian

Kincaid/Clayton
Wills Point / Porters Creek
Matthews Landing
Naheola
Marthaville / Nanafalia
Pendleton / Tuscahoma / Black Mingo / Aquia
Sabinetown Bashi
Hatchetigbee
Reklaw / Tallahatta
Queen City
Weches / Lower Lisbon
Stone City/Dobys Bluff/ Middle Lisbon
Cook Mtn./ Upper Lisbon
Gosport
Moodys Branch
Yazoo

ola
femina
dutexata
apita
mediavia
postmortoni
lisbonensis
arenicola
n.sp. A
mortoni
pleboides
creola
saffordi
chirena
dumblei
alveata
plummeri
clevelandia
kincaidensis
hilli
levicunea
cortezi
rivurbana
stenzeli
subrina
martinensis
tennesseensis
rina
fischeri
alabamiensis
carinata
gilberti
infans
perdita

aldrichi
gardnerae
n.sp. B
humerosa
claytonensis
praecincta
biboraensis
eurynome
toulmini
multilira

3. Larval dispersal was moderate, implying that isolate formation from disturbance (regression) was a main cause of speciation.

4. The level of disturbance was probably moderate, because levels that were too high or too low would not have caused the high origination rates observed.

Fifty-five turritellid species have been identified from the Paleocene and Eocene of the Gulf and Atlantic U.S. Coastal Plain (Allmon 1995). Phylogenetic analysis of these species has employed a combination stratophenetic and cladistic approach (Allmon 1995). This is justified by the relatively high paleontological completeness of the strata (Allmon 1989) and by the relatively small number of discrete shell traits. The results are a phylogenetic tree for 44 species, shown in figure 5.15 (Allmon 1995 provides further details).

The stratigraphic context of the phylogeny indicates that 66% of these gastropod species show their first appearance just after hiatuses associated with regressions (figure 5.15). This pattern results from the actual pattern of origination, combined with three potential biases: stratigraphic, biogeographic, and environmental.

Stratigraphic bias A high proportion of FAD's just after a depositional hiatus may result from originations during the time interval represented by the hiatus. Comparison of patterns among coexisting clades may offer a way to detect correlations between the end of hiatuses and an increase in FAD's because artifactual first appearances would be seen among many taxa, because all would be affected by the same nonpreservation. Differences among taxa in first appearance could be evidence that at least some of the patterns reflect real biological differences that affected speciation rates.

Figure 5.15 Phylogenetic tree of 44 Paleocene and Eocene turritelline species from the U.S. Gulf and Atlantic coastal plains. These constitute 80% of all turritelline species in this time interval and region. *Left:* Composite stratigraphic section of the Coastal Plain Paleogene with major lithostratigraphic units. Major hiatuses marked by disconformities are denoted by heavy black lines (see Toulmin 1977 for stratigraphic details). *Far left:* Sea level curves for the Gulf coast. Eastern Gulf curve modified from Dockery 1986; western Gulf curve modified from Stanton 1982. Figure as a whole modified from Allmon 1995.

Quantitative study by Dockery (1986) of first and last appearances of more than 2,800 species of molluscs from the Coastal Plain has shown that between episodes of high origination and extinction, these faunas also had high background turnover throughout the Paleogene. A high proportion of molluscan species show first appearances in successive intervals during this time (figure 5.16), indicating moderately high origination rates. Overall, the Coastal Plain molluscan fauna thus appears to have generated many new species throughout the Lower Tertiary, not only at certain times or places. A clade-by-clade breakdown clearly needs to be carried out, to see if some clades contribute more or less to this turnover rate. We have attempted two preliminary analyses. Table 5.3 shows first appearance data for all gastropod species in the fauna, broken down by lithostratigraphic units. While some peaks are apparent, they are not striking.

Less ambiguous patterns of first appearances in specific times and environments are available, though rare, in the literature. Two of the best documented examples are shown in figure 5.17. Although further phylogenetic analyses are needed, it is clear that different clades show different first appearance patterns.

We tentatively conclude from all these lines of evidence that because different clades show different patterns of first appearances within the same stratigraphic record, paleontological incompleteness is not the sole cause of those differences. At least some may be due to true differences in speciation.

Biogeographic bias An FAD may record only the local first appearance; the species could have originated elsewhere, and the FAD records the immigration event. Data from the Paleocene and Eocene molluscan faunas of West Africa (Adegoke 1977), Latin America (Allison and Adegoke 1969), California (Merriam 1941), and Europe (Cossman 1912) suggest that turritellid dispersal was occurring then, but detailed evaluation of its magnitude is impaired by lack of recent systematic revisions. Preliminary analysis (Allmon 1995) suggests that no more than 10% of the U.S. Coastal Plain species shown in figure 5.15 originated outside the Coastal Plain.

Environmental bias Transgressions and regressions can produce apparent and real patterns of first and last species appearances (Jablonski

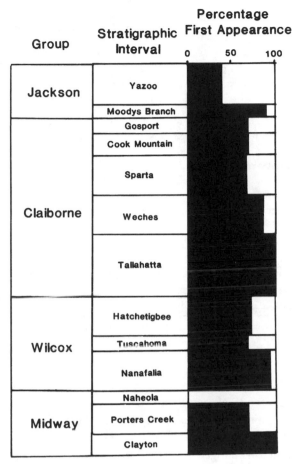

Figure 5.16 Percentages of species in each stratigraphic interval of the Paleocene and Eocene that are recorded as first appearances within the U.S. Gulf Coast. For example, no species in the Tallahatta interval were recorded in previous intervals; in contrast, all species in the Naheola interval are also recorded in a previous interval. Cook Mountain includes Upper Lisbon; Sparta includes Stone City, Middle Lisbon, and Doby's Bluff; Weches includes Lower Lisbon; Tallahatta includes Reklaw; Hatchetigbee includes Bashi and Sabinetown; Tuscahoma includes Pendleton; Nanafalia includes Marthaville; Porters Creek includes Matthews Landing and Wills Point; Clayton includes Kincaid. (See Toulmin 1977 for details of correlation.) Data from Dockery 1986.

Table 5.3
First Appearances of Paleogene Gastropod Species

	Percentages of First Appearances			
Stratigraphic Unit	Texas	Louisiana	Mississippi	Alabama
Moodys Branch			89.3	
Gosport				64.4
Cook Mtn / U Lisbon	32.9	64.2	58.9	26.2
Stone City / Dobys Bluff	58.8		50.0	
Weches / L Lisbon	90.0			89.2
Reklaw / Tallahatta	87.5			
Hatchetigbee				26.2
Sabinetown / Bashi			73.5	
Tuscahoma-Greggs				66.7
Tuscahoma-Bells				32.7
Pendleton	46.9			
Nanafalia-Grampian H.				45.8
Nanafalia-O. *thirsae*				86.9
Matthews Landing				67.2
Wills Pt / Porters Crk	48.8			
Kincaid / Clayton	100.0			100.0

NOTE: First appearance percentages are for gastropod species in the Paleogene Gulf and Atlantic U.S. Coastal Plain, according to strata and state of first recorded appearance.

1980). Since low sea levels are often recorded as a hiatus or nearshore facies, and high sea levels as offshore facies, patterns of first species appearances in certain environments may simply reflect the facies available for study—rather than true origin. But wide-ranging studies can at least partly overcome this danger of misinterpretation. Recent and fossil turritellids show wide ranges of environmental tolerance, occurring in a variety of depths, substrates, temperatures, and current energies (Allmon 1988).

In summary, when all three biases are accounted for and the merely artifactual patterns removed, a majority of Paleogene turritellids are interpreted to have made their first appearance on the Coastal Plain during low sea levels. We can bolster the evidence on the reconstruction of events by adding data on important intrinsic factors to the data on the extrinsic setting just discussed.

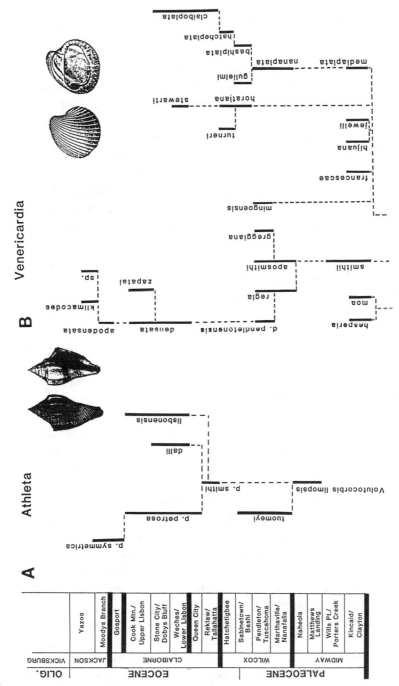

Figure 5.17 Stratophenetically derived phylogenetic trees for two well-studied molluscan lineages from the Gulf Coastal Plain, placed in the same stratigraphic context as figure 5.15. The volutid gastropod *Athleta* is modified from Fisher, Rodda, and Dietrich 1964. The "planicostate" lineages of the carditid bivalve *Venericardia* are modified from Gardner and Bowles 1939.

Peramorphic heterochrony ("overdevelopment" via extension of the ancestral ontogeny) is a major mode of evolution in the turritellids (Allmon 1994). But it is unclear how these heterochronic changes were translated into phylogenetic trends. McNamara (1982) suggested that trends in morphological traits originate from "directed speciation" (cf. Stanley 1979); this occurs along environmental gradients to form heterochronoclines (McKinney and McNamara 1991). A key potential driving force is the production of adaptive novelties that allow the descendant to occupy a new position on the environmental gradient (McNamara 1982).

But it is also possible that much heterochronic change is nonadaptive, being a byproduct of selection on size or other covariant traits (McKinney 1984; McKinney and McNamara 1991). If so, a heterochronocline could be driven not by adaptive novelties but by rate of isolate formation. In turritellids the most important intrinsic factor affecting isolate formation is probably dispersal. Data on developmental mode, geographic range, and stratigraphic duration for turritellids in the present study (Allmon 1989, 1995) indicate that 19 of 40 species shown in figure 5.15 (all those with well-preserved shell apexes) have protoconch whorl numbers and diameters indicative of relatively short planktonic intervals (when interpreted using the values from Jablonski and Lutz 1983). This implies that these species had moderate dispersal abilities, which would not appear to readily promote or hinder isolate formation. (The lower diversity of the *T. humerosa* subclade in figure 5.15 may be related to a longer planktonic period, if Spiller's 1977 report of a 3–4 whorl protoconch is confirmed.)

Figure 5.18 shows a generalized model of how sea level may have affected metapopulation fragmentation to cause speciation and extinction. Increasing regression enhances the probability (rate) of isolate formation, as local topographic irregularities cause semi-isolation of small inlets and bays. Probability (rate) of isolate persistence would tend to diminish in increasingly stressed environments, such as increased turbidity, loss of shelf area, and crowding. Net speciation would thus tend to be low (and extinction increasing) from the onset of regression until the regressive maximum is reached.

As transgression begins anew, the probability (rate) of isolate persistence would tend to increase, as environmental conditions improve. Maximum speciation rates might be expected when the transgression is in its intermediate stages, as isolate persistence would be relatively high and the rate of isolate formation is also higher than it would be when peak trans-

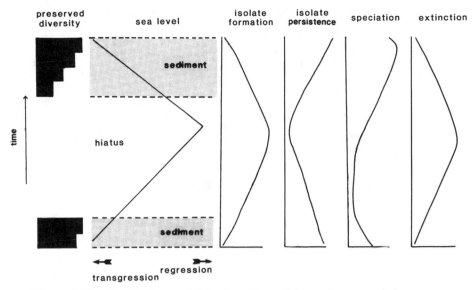

Figure 5.18 Qualitative model for the effects of the environmental change during an episode of regression followed by transgression on rates of isolate formation, isolate persistence, speciation, and extinction of established species in benthic marine invertebrates. Each of these four kinds of organic changes is greatest at the time(s) that its line is farthest to the right. The regression creates an unconformity, and thus most of the biological changes happen when no permanent stratigraphic record is being deposited.

gression is reached (figure 5.18). In accordance with our intermediate disturbance model, we thus expect maximum species origination, and perhaps moderate extinction, to coincide with intermediate levels of transgression. This prediction is corroborated by other empirical evidence, such as that of Jablonski (1980) and Kauffman (1973), who indicate that speciation is maximized during transgression, owing to enhanced endemism. Our model deviates slightly from their interpretations in our emphasis on the intermediate stages, rather than the peak, of the transgressive episode as being the time of maximum speciation.

Implications for Onshore Origination, Life History, and Ecology

Any complete discussion of speciation in the marine fossil record must address the widely discussed pattern of onshore origin, which is the tendency for higher taxa or major novelties to originate in peritidal to shallow water environments (Jablonski and Bottjer 1990). Many (largely untested) theories have been advanced to account for this origination pattern (review in Sepkoski 1991). Our intermediate disturbance model would imply that any explanation should incorporate the likelihood that environments of moderate disturbance are involved. An explanation should also incorporate the observations of Parsons (1991, 1993) that moderate disturbance is most likely, and most effective for reasons of intrinsic variation, near the species range boundaries.

Figure 5.19 (top) shows that increasing intensity and frequency of environmental disturbance can serve as a primary vector of extrinsic control on many biological patterns, including origination, extinction, and diversity. Figure 5.19 (bottom) presents a model that relates this vector to water depth (*sensu* McKinney 1984, 1986; see also Sepkoski 1987 which finds increasing extinction rate in an onshore direction). According to Parsons (1991, 1993), there are many reasons that maximum speciation rates should occur in areas of intermediate disturbance within a species range. Among the most important is that very high disturbance requires an organism to possess stress adaptations with high metabolic costs, whereas very low disturbance tends to minimize genetic and phenotypic variability. Figure 5.19 (bottom) therefore suggests that the effects of intermediate disturbance are most pronounced in species adapted to relatively stable environments.

Sheldon (1993) has suggested that the greater effect of intermediate disturbance on stable-adapted species occurs because, unlike "generalist" opportunistic species, species adapted to less dynamic environments tend to be more sensitive to environmental change. This sensitivity, in turn, leads to higher speciation rates. This argument agrees with many traditional explanations we have already discussed—namely, that more specialized, stenotopic species tend to have higher speciation rates. Not only are they more narrowly adapted physiologically but they also may be morphologically or behaviorally specialized, with "more parts to change" (McKinney and McNamara 1991). Vermeij (1973) called this the "law of independent parameters": more parameters mean more kinds of change is possible, and it often means greater sensitivity to environmental change. An example may be the higher turnover shown by fine-filter pinnulate crinoids than by coarse-filter non-pinnulates (Baumiller 1993).

Our tentative model (figure 5.19, bottom) also suggests the cause of

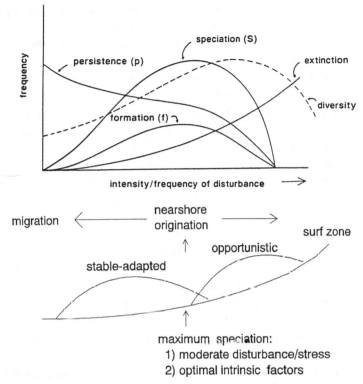

Figure 5.19 *Bottom:* A model depicting how spatial aspects of species distributions (from nearshore to surf zone, or migration from/to other locales) and ecology (stable-adapted v. opportunistic) may affect speciation. High rates of origination ("speciation" and "formation" in the top graph) occur nearshore (1) where there are intermediate levels of disturbance (the extrinsic factor promoting speciation) and (2) if the location is near the fringes of species ranges, where maximum variability occurs, and other intrinsic factors promoting speciation (see text) are present. Subsequent migration offshore leads to an onshore-offshore pattern.

the onshore origination of taxa and their subsequent offshore migration. Jablonski and Bottjer (1990) have rejected McKinney's (1986, based on Gould 1977) earlier model, which is based on preferential origination by paedomorphosis (especially progenesis). McKinney (McKinney and McNamara 1991) has now also rejected his early model, and for numerous reasons. Most relevant here is that de Beer's concept that paedomorphs have more evolutionary potential is tautological and unproven. Indeed, theory suggests that peramorphs may have more potential because specializations may open new areas of niche space (Levinton 1988; McKinney and McNamara 1991). This contention pertaining to niche creation agrees with our model's suggestion that less generalized species in more stable environments not only speciate more often but have sufficient evolutionary potential for ordinal originations. That the ordinal radiation later migrates offshore (and nearshore representatives become rarer) may also be explained by our model. Offshore migration would be expected if the original ordinal ancestors were from taxa adapted to stabler environments. Subsequent rarity would be expected if the nearshore "speciation pump" created new taxa that supplanted the older forms. For those species adapted to stabler environments, competition and other biotic interactions would be major factors in controlling the abundance of populations that had migrated further offshore.

Our model agrees, in many ways, with the interesting explanation by Sepkoski (1991) of onshore-offshore patterns. His model showed that extinction resistance alone could produce the patterns when more extinction-resistant taxa moved up the disturbance (extinction) gradient toward nearshore and then diversified there. Increasingly resistant taxa would replace less resistant ones, which would tend to persist mainly in less disturbed offshore environments. Our intermediate disturbance model conforms to Sepkoski's in having the focus of origination outside the zone of highest disturbance, occurring in stable-adapted species which then migrate up the disturbance gradient. Later cycles of replacement by more extinction-resistant clades would leave unaffected only those that migrated offshore. Although Sepkoski's (1991) model uses extinction resistance to determine the long-term patterns of replacement, origination obviously still must occur; but it is treated as a constant. We are simply suggesting here that whatever origination occurs must have a geographical and an ecological locus, and that it must account for species variation and other intrinsic properties.

Relationship of the Intermediate Disturbance Model to r-K Theory

Obviously, our intermediate disturbance model is very tentative and very coarse. Even so, it serves as a testable null hypothesis. Despite the continued use of r-K theory by some, we agree with two recent major reviews (Roff 1992; Stearns 1992) that r-K theory is an oversimplification. Life history traits do not covary in the uniform way once thought, at least at the ecological (population) scales for which it was formulated. At higher phylogenetic levels, however, such covariation does exist. For example, insects are "r-selected" relative to mammals. Such covariation is not limited to phylum level comparisons but can be seen at class, order, and family levels. Stearns (1992) notes that these patterns represent life history "decisions" made long ago which now reflect constraints.

To ecologists, these coarse phylogenetic "mouse-to-elephant" patterns have served mainly as a troublesome source of confusion because they were conflated with ecological patterns (e.g., Bonner 1988). But given the coarse scales viewed by the paleobiologist, such coarse patterns may be useful. For example, we might infer from our model that certain families or orders that are relatively "r-selected," inhabiting mainly nearshore and other often disturbed environments, would tend to speciate less often. Indeed, Jablonski and Bottjer (1990) examined clade origination from such higher clade levels.

More directly relevant to the ecological scale of figure 5.19 (bottom) is that some key aspects of r-K still seem viable at the population level. For example, Stearns (1992) discusses how high juvenile mortality can lead to delayed maturation and slower growth. Such a selective regime is likely to be found in a relatively stable environment where competition is a major factor. This is especially true where competitors include juveniles and adults of other species, as described by the ontogenetic niche concept (Werner and Gilliam 1984). Conversely, Roff (1992) notes how environments with undersaturated carrying capacities place a selective premium on rapid reproductive rate; the life history trait of early maturation would be an advantage in this circumstance.

Highly disturbed environments will obviously tend to be the most undersaturated. The high juvenile and adult mortality of disturbed areas also favor early maturation. For instance, the "opportunistic" species concept, found in many aspects of ecological theory, has withstood many tests and is a fundamental part of community succession. Parsons (1993) incorporates the concept into a general theory addressing stress, metabolic energy, and disturbance. Indeed, his views are part of our disturbance model. Stearn's (1992) review concludes that relating habitat to life history may be the most promising approach (e.g., Southwood 1988). In the present context, this might mean that mortality patterns of disturbed versus undisturbed habitats may hold the key to explaining the type of populational patterns suggested in figure 5.19.

If nothing else, this testable null model we present illustrates how fine-scale analyses of fossil populations can contribute badly needed evidence for construction of a more complex life history model to replace r-K, thus relating population dynamics to coarser biodiversity patterns in the record. For example, 63% of the 16 Late Eocene echinoid species in a shallow water environment co-occur with one or more other species of the same genus. In contrast, only 43% of the 21 echinoid species in deeper waters co-occur with another species of the same genus (data in Carter and McKinney 1992, appendixes IC, IG). Assuming that taxonomic (and morphological) similarity approximates niche proximity, the echinoid data seems to agree with the notion that competition is more important in determining which species inhabit deeper, stabler environments. Conversely, abiotic disturbance is implicated as more important in structuring communities in shallower environments, an assumption that is fundamental to Sepkoski's (1991) model.

Summary: Scaling Up Metapopulation Dynamics

Probably all species persist as metapopulations at an appropriate
spatial scale.
PAUL HARVEY AND COLLEAGUES, 1993

The fossil record can be fruitfully studied as the product of more than 3.5 billion years of metapopulation dynamics. But paleontologists have not devoted much attention to studying population-level processes in the record. We believe that neglect of population-level processes owes to an overemphasis on the poverty of the record, combined with inadequate theoretical models and conceptual frameworks for how speciation occurs and how it may be studied in the record.

We have tried to overcome some of these inadequacies by applying the growing and conceptually rich field of metapopulation dynamics, where a species is analyzed as patches of populations of varying size and duration. The fossil record obviously does not record the dynamics of all metapopulation patches at all times. But we would also argue that at least some data on patchiness, especially spatial aspects of abundance and morphological variation, are often available if we choose to collect it. Whether this data reflects fine or coarse scales of change will depend on the conditions and taxa observed, but the wealth of possible research tools offered by metapopulation theory provides a framework for interpreting the data. For example, by focusing on the formation, persistence, and differentiation of patches, a series of reasonable and (most importantly) testable predictions can be made about speciation and extinction in both regional (ecological) and taxonomic (phylogenetic) contexts.

At a fine scale, population dynamics can be studied in the record by reference to what we term the "particle-patch model," wherein a stratigraphic section is seen as a point in space across which patches migrate through time. Very patchy metapopulations, with small or low-density patches, would be recorded as having highly variable abundances upsection (i.e., spatial patchiness translates into vertical patchiness). Geographic range of metapopulations would be visible as the duration that the species persists in the section (the time it takes to migrate away from the depositional area). Endemic species would be expected to have relatively low mean abundance, high abundance variation, and short vertical duration. Such fine-scale metapopulation characteristics as peripheral "sink" populations may be identified by increased abundance variation ("red shifts"), increased morphological variation from stress, and other clues. This fine-scale patch information can provide invaluable evidence on speciation dynamics when coupled with other evidence, such as species richness. For example, do patchier species tend to speciate more often? Are they more widespread?

At coarser scales, patterns of first and last appearance can be meaningful evolutionary data, especially when carefully combined with geographic, taxonomic, and taphonomic information. Data on dispersal and selective pressures can be applied to answer such questions as whether population isolate persistence and isolate formation were the primary controls on speciation.

A major inference of the metapopulation approach is that disturbance, i.e., environmental change, plays a basic role in both speciation and extinction. This inference has allowed us to develop an analytical framework to examine the empirical linkage between speciation and extinction, visible through time both within taxa (clades) and over geographic areas. A linkage exists because patch fragmentation plays a key role in both origination and extinction at the species level. We propose an *intermediate disturbance model of maximal speciation* for the origin of biodiversity through evolutionary time. At low levels of disturbance, speciation rates are low because little vicariance and fragmentation occurs. At high levels of disturbance, speciation rates are also low because isolate persistence is diminished, owing to high stress. Moderate levels of both isolate formation and persistence are, however, present at levels of intermediate disturbance.

The predictions of our intermediate disturbance model of origination can be applied to all levels of the biological hierarchy, from the population patch to higher taxa (clades), and in the ecological realm, from local communities (alpha diversity) to the region and biosphere (gamma and delta diversity). This is because disturbance occurs at all levels and has the same effect of fragmentation at all scales, varying in extent (geographic scope) and intensity (magnitude or number of environmental parameters affected).

This property of fragmentation effects across many spatial scales is consistent with the fractal concept of scale-free self-similar behaviors. That so many aspects of speciation and extinction—from disturbance magnitude to patch sizes to genus durations—follow the hollow curve (power law) pattern indicative of fractal processes may thus not be coincidental. Rather, it implies that fundamental laws of disturbance and fragmentation may control many of the origination and extinction patterns we see in the fossil record. "Secondary" dynamics (e.g., historical and taxon-specific factors) would undoubtedly modify these fundamental laws. But we suggest that a deeper understanding of biodiversity turnover would result from consideration of the ideas proposed in this paper. Perhaps one of the most important insights is that studying either origination or extinction by itself is much less informative than studying them jointly. A complete understanding of speciation would include the dynamics of biodiversity turnover at all spatiotemporal scales (McKinney and Drake, in press).

Acknowledgments   Contributions by Michael L. McKinney were supported by the National Science Foundation (NSF EAR-931617) and the American Chemical Society (ACS-PRF22635-AC8).

References

Adegoke, O. 1977. Stratigraphy and paleontology of the Ewekoro Formation (Paleocene) of southwestern Nigeria. *Bulletins of American Paleontology* 71:1–379.

Allison, P. and D. Briggs. 1991. *Taphonomy: Releasing the Information in the Fossil Record*. New York: Plenum.

Allison, R. and O. Adegoke. 1969. The *Turritella rina* group (Gastropoda) and its relationship to *Torcula* Gray. *Journal of Paleontology* 43:1248–1266.

Allmon, W. D. 1988. Ecology of living turritelline gastropods (Prosobranchia, Turritellidae): Current knowledge and paleontological implications. *Palaios* 3:259–284.

———. 1989. Paleontological completeness of the record of Paleogene molluscs, U.S. Gulf and Atlantic Coastal Plains: Implications for phylogenetic studies. *Historical Biology* 3:141–158.

———. 1992. A causal analysis of stages in allopatric speciation. *Oxford Surveys in Evolutionary Biology* 8:219–257.

———. 1994. Patterns and processes of heterochrony in lower Tertiary turritelline gastropods, U.S. Gulf and Atlantic Coastal Plains. *Journal of Paleontology* 68:80–95.

———. 1995. Evolution and systematics of Cenozoic American Turritellidae (Gastropoda). I. Paleocene and Eocene species related to "Turritella mortoni Congrad" from the U.S. Gulf and Atlantic coastal plains. *Palaeontographica Americana*. In press.

Allmon, W. D., G. Rosenberg, R. Portell, and K. Schindler. 1993. Diversity of Atlantic Coastal Plain mollusks since the Pliocene. *Science* 260:1626–1629.

Baumiller, T. K. 1993. Survivorship analysis of Paleozoic Crinoidea: Effect of filter morphology on evolutionary rates. *Paleobiology* 19:304–321.

Berg, H. 1981. *Random Walks in Biology*. Princeton: Princeton University Press.

Bonner, J. T. 1988. *The Evolution of Complexity*. Princeton: Princeton University Press.

Boucot, A. J. 1975. *Evolution and Extinction Rate Controls*. Amsterdam: Elsevier.

Brooks, D. R. 1991. Neontology and Paleontology. *Trends in Ecology and Evolution* 6:198–199.

Burlando, B. 1990. The fractal dimension of taxonomic systems. *Journal of Theoretical Biology* 146:99–114.

Carter, B. and M. McKinney. 1992. Eocene echinoids, the Suwannee Strait, and biogeographic taphonomy. *Paleobiology* 18:299–325.

Charlesworth, B., R. Lande, and M. Slatkin. 1982. A neo-Darwinian commentary on macroevolution. *Evolution* 36:474–498.

Cossman, A. 1912. *Essais de Paleoconchologie Comparee*. Paris. Privately published.

Cracraft, J. 1992. Explaining patterns of biological diversity. In N. Eldredge, ed., *Systematics, Ecology, and the Biodiversity Crisis*, pp. 59–76. New York: Columbia University Press.

Dockery, D. T. 1986. Punctuated succession of Paleogene molluscs in the northern Gulf Coastal Plain. *Palaios* 1:582–589.

Donovan, S. K., ed., 1989. *Mass Extinctions*. New York: Columbia University Press.

Dott, R. H. 1983. Episodic sedimentation—how normal is normal? *Journal of Sedimentary Petrology* 53:5–23.

Eldredge, N. 1992. Where the twain meet: Causal intersections between the genealogical and ecological realms. In N. Eldredge, ed., *Systematics, Ecology, and the Biodiversity Crisis*, pp. 1–14. New York: Columbia University Press.

Eldredge, N. and S. J. Gould. 1972. Punctuated equilibria: An alternative to phyletic gradualism. In T. J. M. Schopf, ed., *Models in Paleobiology*, pp. 82–115. San Francisco: Freeman Cooper.

Endler, J. A. 1989. Conceptual and other problems in speciation. In D. Otte and J. A. Endler, eds., *Speciation and its Consequences*, pp. 625–650. Sunderland, Mass.: Sinauer.

Erickson, R. 1945. The *Clematis fremontii* population in the Ozarks. *Annals of the Missouri Botanical Garden* 32:413–460.

Erwin, D. 1993. *The Great Paleozoic Crisis*. New York: Columbia University Press.

Fischer, W., P. Rodda, and J. Dietrich. 1964. Evolution of *Athleta petrosa* stock (Eocene, Gastropoda) of Texas. *University of Texas Publication* 6413.

Flessa, K. W. and R. H. Thomas. 1985. Modeling the biogeographic regulation of evolutionary rates. In J. Valentine, ed., *Phanerozoic Diversity Patterns*, pp. 355–373. Princeton: Princeton University Press.

Flessa, K. W. and Group Report. 1986. Causes and consequences of extinction. In D. Raup and D. Jablonski, eds., *Patterns and Processes in the History of Life*. Berlin: Springer-Verlag.

Frontier, S. 1986. Applications of fractal theory to ecology. In P. Legendre and L. Legendre, eds., *Developments in Numerical Ecology*, pp. 335–380. Berlin: Springer-Verlag.

Fursich, T. and M. Aberhan. 1990. Significance of time-averaging for paleocommunity analysis. *Lethaia* 23:143–152.

Gardner, J. and E. Bowles. 1939. The *Venericardia planicosta* group in the Gulf province. *United States Geological Survey Professional Paper* 189:141–215.

Gingerich, P. D. 1985. Species in the fossil record: Concepts, trends, and transitions. *Paleobiology* 11:27–41.

Gould, S. J. 1977. *Ontogeny and Phylogeny*. Cambridge: Harvard University Press.

———. 1982. The meaning of punctuated equilibrium and its role in validating a hierarchical approach to macroevolution. In R. Milkman, ed., *Perspectives on Evolution*, pp. 83–104. Sunderland, Mass.: Sinauer.

————. 1985. The paradox of the first tier: An agenda for paleobiology. *Paleobiology* 11:2–12.

————. 1989. A developmental constraint in *Cerion*. *Evolution* 43:516–539.

Hallam, A. 1972. Models involving population dynamics. In T. J. M. Schopf, ed. *Models in Paleobiology*, pp. 62–81. San Francisco: Freeman Cooper.

Hansen, T. 1988. Early Tertiary radiation of marine molluscs and the long-term effects of the K-T extinction. *Paleobiology* 14:37–51.

Hanski, I. 1989. Metapopulation dynamics: Does it help to have more of the same? *Trends in Ecology and Evolution* 4:113–114.

Hanski, I. and M. Gilpin. 1991. Metapopulation dynamics: Brief history and conceptual domain. *Biological Journal of the Linnean Society* 42: 3–16.

Harrison, S. and J. F. Quinn. 1989. Correlated environments and the persistence of metapopulations. *Oikos* 56:293–298.

Harvey, P., S. Nee, A. Mooers, and L. Partridge. 1993. These hierarchical views of life: Phylogenies and metapopulations. In T. Crawford and E. Hewitt, eds., *Genes in Evolution*, pp. 123–137. Oxford: Blackwell.

Hastings, A. and C. Wolin. 1989. Within-patch dynamics in a metapopulation. *Ecology* 70:1261–1266.

Hengeveld, R. 1992. *Dynamic Biogeography*. Cambridge: Cambridge University Press.

Hengeveld, R. and J. Haeck. 1982. The distribution of abundance. *Journal of Biogeography* 9:303–316.

Hoffman, A. 1989. *Arguments on Evolution*. Oxford: Oxford University Press.

Hsu, K. J. 1989. Catastrophic extinctions and the inevitability of the improbable. *Journal of the Geological Society of London* 146:749–754.

Jablonski, D. 1980. Apparent versus real biotic effects of transgressions and regressions. *Paleobiology* 6:397–407.

————. 1989. The biology of mass extinction: A paleontological view. *Philosophical Transactions of the Royal Society of London* B325:357–368.

Jablonski, D. and D. Bottjer. 1990. Onshore-offshore trends in marine invertebrate evolution. In R. Ross and W. D. Allmon, eds., *Causes of Evolution: A Paleontological Perspective*, pp. 21–75. Chicago: University of Chicago Press.

Jablonski, D. and R. Lutz. 1983. Larval ecology of marine benthic invertebrates: Paleobiological implications. *Biological Reviews* 58:21–89.

Jackson, J. B. C. 1974. Biogeographic consequences of eurytopy and stenotopy among marine bivalves and their evolutionary significance. *American Naturalist* 108:541–560.

Jackson, J. B. C., P. Jung, A. Coates, and L. Collins. 1993. Diversity and extinction of tropical American mollusks and emergence of the Isthmus of Panama. *Science* 260:1624–1626.

Johnson, G. A. 1982. Occurrence of phyletic gradualism and punctuated equilibria through geologic time. *Journal of Paleontology* 52:1329–1331.

Kauffman, E. G. 1973. Cretaceous bivalvia. In A. Hallam, ed., *Atlas of Paleobiogeography*, pp. 353–384. Amsterdam: Elsevier.

————. 1977. Evolutionary rates and biostratigraphy. In E. Kauffman and J. Hazel, eds., *Concepts and Methods of Biostratigraphy*, pp. 109–142. Stroudsburg: Dowden, Hutchinson and Ross.

Kidwell, S. 1993. Patterns of time averaging in the shallow marine fossil record. In S. Kidwell and A. Behrensmeyer, eds., *Taphonomic Approaches to Time Resolution in Fossil Assemblages*, pp. 275–300. Knoxville: Paleontological Society.

Kidwell, S. and A. Behrensmeyer, eds. 1993. *Taphonomic Approaches to Time Resolution in Fossil Assemblages*. Knoxville: Paleontological Society.

Kidwell, S., and D. Bosence. 1991. Taphonomy and time-averaging of marine shelly faunas. In P. Allison and D. Briggs, eds., *Taphonomy: Releasing the Data Locked in the Fossil Record*, pp. 115–209. New York: Plenum.

Koch, C. 1987. Prediction of sample size effects on the measured temporal and geographic distribution patterns of species. *Paleobiology* 13:100–107.

————. 1991. Sampling from the fossil record. In N. Gilinsky and P. Signor, eds., *Analytical Paleobiology*, pp. 4–18. Knoxville: Paleontological Society.

Kunin, W. E. and K. Gaston. 1993. The biology of rarity: Patterns causes and consequences. *Trends in Ecology and Evolution* 8:298–301.

Lande, R. 1981. The minimum number of genes contributing to quantitative variation between and within populations. *Genetics* 99:541–553.

————. 1986. The dynamics of peak shifts and the pattern of morphological evolution. *Paleobiology* 12:343–354.

Lande, R. and G. Barrowclough. 1987. Effective population size genetic variation, and their use in population management. In M. Soule, ed., *Viable Populations for Conservation*, pp. 87–123. Cambridge: Cambridge University Press.

Levins, R. 1969. Some demographic and genetic consequences of environmental heterogeneity for biological control. *Bulletin of the Entomological Society of America* 15:237–240.

Levinton, J. S. 1988. *Genetics, Paleontology, and Macroevolution*. Cambridge: Cambridge University Press.

Magurran, A. 1988. *Ecological Diversity and Its Measurement*. Princeton: Princeton University Press.

Marshall, C. 1991. Estimation of taxonomic ranges from the fossil record. In N. Gilinsky and P. Signor, eds., *Analytical Paleobiology*, pp. 19–38. Knoxville: Paleontological Society.

Martin, R. 1993. Time and taphonomy: Actualistic evidence for time-averaging of benthic foraminiferal assemblages. In S. Kidwell and A. Behrensmeyer, eds., *Taphonomic Approaches to Time Resolution in Fossil Assemblages*, pp. 34–56. Knoxville: Paleontological Society.

McKinney, M. L. 1984. Allometry and heterochrony in an Eocene echinoid lineage: Morphological change as a byproduct of size selection. *Paleobiology* 10:407–419.

————. 1986. Ecological causation of heterochrony: A test and implications for evolutionary theory. *Paleobiology* 12:282–289.

McKinney, M. L. and J. A. Drake, eds. In press. *Biodiversity Dynamics*. New York: Columbia University Press.

McKinney, M. L. and D. Frederick. 1992. Extinction and population dynamics: New methods and evidence from Paleogene foraminifera. *Geology* 20:343–346.

————. submitted. Migration of population patches across stratigraphic sections. *Palaios*.

McKinney, M. L. and K. J. McNamara. 1991. *Heterochrony: The Evolution of Ontogeny*. New York: Plenum.

MacLeod, N. 1991. Punctuated anagenesis and the importance of stratigraphy to paleobiology. *Paleobiology* 17:167–188.

McNamara, K. J. 1982. Heterochrony and phylogenetic trends. *Paleobiology* 8:130–142.

McNaughton, S. J. and L. L. Wolf. 1970. Dominance and the niche in ecological systems. *Science* 167:131–139.

Merriam, C. W. 1941. Fossil turritellas from the Pacific Coast region of North America. *University of California Publications in Geological Sciences Bulletin* 26.

Miller, A. I. 1988. Spatial resolution in subfossil molluscan remains: Implications for paleobiological analyses. *Paleobiology* 14:91–103.

Murphy, D., K. Freas, and S. Weiss. 1990. An environment-metapopulation approach to population viability analysis for a threatened invertebrate. *Conservation Biology* 4:41–51.

Murray, J. 1991. *Ecology and Paleoecology of Benthic Foraminifera*. New York: Wiley.

Murray, J. D. 1988. Spatial dispersal of species. *Trends in Ecology and Evolution* 3:307–309.

Otte, D. and J. A. Endler. 1989. *Speciation and Its Consequences*. Sunderland, Mass.: Sinauer.

Parsons, P. A. 1991. Evolutionary rates: Stress and species boundaries. *Annual Review of Ecology and Systematics* 22:1–16.

————. 1993. Stress, extinctions, and evolutionary change: From living organisms to fossils. *Biological Reviews* 68:313–333.

Petraitis, P. S., R. Lathan, and R. Niesenbaum. 1989. The maintenance of species diversity by disturbance. *Quarterly Review of Biology* 64:393–418.

Pimm, S. 1991. *The Balance of Nature?* Chicago: University of Chicago Press.

Pimm, S. and A. Redfearn. 1988. The variability of animal populations. *Nature* 334:613–614.

Plotnick, R. and M. L. McKinney. 1993. Ecosystem organization and extinction dynamics. *Palaios* 8:202–212.

Prothero, D. and W. Berggren. 1992. *Eocene-Oligocene Climatic and Biotic Evolution*. Princeton: Princeton University Press.

Pulliam, H. R. 1988. Sources, sinks, and population regulation. *American Naturalist* 132:652–661.

Rabinowitz, D., S. Cairns, and T. Dillon. 1986. Seven forms of rarity and their frequency in the flora of the British Isles. In M. Soule, ed., *Conservation Biology*, pp. 182–204. Sunderland, Mass.: Sinauer.

Raup, D. M. 1991a. *Extinction: Bad Genes or Bad Luck?* New York: W. W. Norton.

———. 1991b. A kill curve for Phanerozoic marine species. *Paleobiology* 17:37–48.

———. 1992. Large-body impacts and extinction in the Phanerozoic. *Paleobiology* 18:80–88.

Ricklefs, R. 1990. *Ecology*. New York: W. H. Freeman.

Roff, D. A. 1992. *The Evolution of Life Histories: Theory and Analysis*. London: Chapman and Hall.

Sadler, P. and D. Strauss. 1990. Estimation of completeness of stratigraphical sections using empirical data and theoretical models. *Journal of the Geological Society of London* 147:471–485.

Saul, L. 1983. *Turritella* zonation across the Cretaceous-Tertiary boundary. *University of California Publications in Geological Sciences* 125:1–164.

Schroeder, M. 1991. *Fractals, Chaos, and Power Laws*. New York: W. H. Freeman.

Sepkoski, J. J. Jr. 1987. Environmental trends in extinction during the Paleozoic. *Science* 235:64–66.

———. 1991. A model of onshore-offshore change in faunal diversity. *Paleobiology* 17:58–77.

Sheldon, P. R. 1987. Parallel gradualistic evolution of Ordovician trilobites. *Nature* 330:561–563.

———. 1993. Making sense of microevolutionary patterns. In D. R. Lees and D. Edwards, eds., *Evolutionary Patterns and Processes. Linnean Society Symposium* 14:19–31. London: Academic Press.

Signor, P. W. 1990. The geologic history of diversity. *Annual Review of Ecology and Systematics* 21:509–539.

Southwood, T. R. 1988. Tactics, strategies, and templets. *Oikos* 52:3–18.

Spiller, J. 1977. Evolution of turritellid gastropods from the Miocene and Pliocene of the Atlantic Coastal Plain. Ph.D. dissertation. State University of New York, Stonybrook.

Stanley, S. M. 1979. *Macroevolution: Pattern and Process*. San Francisco: W. H. Freeman.

———. 1982. Macroevolution and the fossil record. *Evolution* 36:460–473.

———. 1986. Population size, extinction, and speciation: The fission effect in Neogene Bivalvia. *Paleobiology* 12:89–110.

———. 1990. The general correlation between rate of speciation and rate of extinction: Fortuitous causal linkages. In R. Ross and W. D. Allmon, eds., *Causes of Evolution: A Paleontological Perspective*, pp. 103–127. Chicago: University of Chicago Press.

Stanton, R. J. 1982. Cenozoic stratigraphy of Texas. In R. Maddocks, ed., *Texas Ostracodes. Guidebook of Excursions for the Eighth International Symposium on Ostracods.* Houston: University of Houston.

Stearns, S. C. 1992. *The Evolution of Life Histories.* Oxford: Oxford University Press.

Steele, J. 1986. A comparison of terrestrial and marine ecological systems. *Nature* 313:355–358.

Toulmin, L. 1977. Stratigraphic distribution of Paleocene and Eocene fossils in the eastern Gulf Coast region. *Alabama Geological Survey Monograph* 13.

Underwood, A. J. 1989. The analysis of stress in natural populations. In P. Calow and R. Berry, eds., *Evolution, Ecology and Environmental Stress,* pp. 51–78. London: Academic Press.

Vermeij, G. J. 1973. Morphological patterns in high intertidal gastropods: Adaptive strategies and their limitations. *Marine Biology* 20:319–346.

Vrba, E. S. 1987. Ecology in relation to speciation rates: Some case histories of Miocene-Recent mammal clades. *Evolutionary Ecology* 1:283–300.

Werner, E. E. and J. Gilliam. 1984. The ontogenetic niche and species interactions in size-structured populations. *Annual Review of Ecology and Systematics* 15:393–425.

West, B. J. and Shlesinger, M. 1990. The noise in natural phenomena. *American Scientist* 78:40–45.

Whitlock, M. 1992. Nonequilibrium population structure in forked fungus beetles. *American Naturalist* 139:952–970.

Williamson, P. G. 1987. Selection or constraint? A proposal on the mechanism of stasis. In K. Campbell and M. Day, eds., *Rates of Evolution,* pp. 129–142. London: Allen and Unwin.

6

Process from Pattern:
Tests for Selection Versus Random Change
in Punctuated Bryozoan Speciation

Alan H. Cheetham and Jeremy B. C. Jackson

The tempo of speciation in marine invertebrates has become increasingly clear through detailed analyses of morphologic evolution in well-defined, geographically widespread fossil lineages. Phenotypic differentiation of some species is gradual, extending over intervals of 10^5 to 10^6 years, with clear records of intermediate morphologies (Malmgren and Kennett 1981; Reyment 1985; Geary 1991, 1992, and this volume). Many other species appear abruptly in the stratigraphic record, commonly but not exclusively without known intermediates, and persist with little or no change for periods of 10^6 years or longer (Cheetham 1986; Wei and Kennett 1988; Geary 1991, 1992, and this volume). Phenotypic divergence of species with static morphologies is apparently concentrated in geologically brief periods of speciation whose upper limits are set by biostratigraphic precision. Precision depends on variation in sediment accumulation rate (MacLeod 1991), but it is typically on the order of 10^4 to 10^5 years.

Despite clarification of the patterns of phenotypic change, interpretation of the underlying modes of speciation remains controversial, at least partly because genetic models that emphasize different factors of population size and degrees of geographic isolation are plausible and not mutually exclusive (Barton and Charlesworth 1984). In addition, genetic data supporting these models are largely from groups, such as birds on islands and insects on plants, with insufficiently detailed fossil records to interpret in long-term historical perspective (Otte and Endler 1989).

How different are the processes of speciation underlying gradual and punctuated patterns? In particular, do differences in pattern imply differences in the relative importance of selection (both directional and stabilizing) and random genetic change (mutation and genetic drift)? Directional selection is commonly invoked to explain gradual differentiation of species, but the level of selection required for trends of net change sustained over intervals of 10^6 years or more may be so weak that random change may be as likely an explanation (Geary 1991), unless every fluctuation in morphology was brought about by environmental change. Punctuated patterns of speciation are consistent with a variety of genetic models that involve shifts between adaptive peaks on which phenotypes remain relatively static through long-term stabilizing selection (Wright 1982; Lande 1985, 1987; Lynch 1990). The geologically abrupt divergence of species in such models, on time scales of 10^5 years or less, seems more likely to involve directional selection. However, Lynch (1990) has shown that rates of phenotypic divergence for a wide variety of mammals, the taxa most often cited as examples of rapid evolution, could have resulted entirely from the random processes of mutation and genetic drift interacting with stabilizing selection.

In addition to the problem of how closely evolutionary tempo may reflect speciation mode, the phenotypic patterns themselves must meet challenges on at least three grounds. (1) To what extent do phenotypic changes preserved in the fossil record correspond to differences between genetically distinct, biologic species (Turner 1986; Levinton 1988)? Morphospecies based on preservable skeletal morphology could instead represent either suites of cryptic species (Larson 1989) or intraspecific morphologic variants (Kat and Davis 1983; Palmer 1985). (2) With what precision do stratigraphic first appearances provide evidence for the timing and geographic location of species origins (Marshall 1990, 1991, and this volume; MacLeod 1991)? Abrupt appearances in the record could instead result from preservation failure at major hiatuses, time-averaging in condensed sections, or migration from unsampled areas. (3) Given the imperfection of the fossil record, how well does the nearest-neighbor, stratophenetic approach reconstruct phylogenetic relationships among species, and how do the results compare with a cladistic approach?

Here we review patterns of phenotypic evolution for two Tertiary-to-Recent genera of cheilostome bryozoans, *Metrarabdotos* and *Stylopoma*, and explore their possible modes of speciation. *Metrarabdotos* is one of the taxa from which the most persuasive evidence for punctuated speciation has come (Gould 1991), and the less detailed record of *Stylopoma* appears to conform to the same pattern (Jackson and Cheetham 1994). To obtain a first approximation of the roles of selection and random change in these genera, we employ a method of analysis developed by Lynch (1990) based on quantitative genetics, but requiring only phenotypic data directly available for fossil as well as living populations.

Patterns of Phenotypic Evolution in *Metrarabdotos* and *Stylopoma*

Metrarabdotos (figure 6.1) and *Stylopoma* (figure 6.2) are ascophoran cheilostomes, both of which underwent significant radiation in the Neogene and Quaternary of tropical America (Cheetham 1968, 1986; Jackson and Cheetham 1994; Cheetham and Jackson 1995). Morphospecies were discriminated in both genera by the same series of morphometric and statistical procedures to provide as uniform a basis as possible for the analysis of phenotypic evolution (Cheetham 1986; Jackson and Cheetham 1990, 1994). To maximize the sensitivity of morphologic discrimination, morphospecies were distinguished on the smallest statistically significant intercolony differences in overall zooidal morphology, with replicate zooidal measurements within colonies providing a check against oversplitting. Despite large differences in numbers of morphologic characters used (46 in *Metrarabdotos* versus 12 in *Stylopoma*) and in average morphologic distances between species, the number of morphospecies distinguished in each genus (19) is the same (table 6.1). However, morphologic distances also vary considerably between subsets of species within each genus (table 6.1).

The genetic basis of morphospecies has been investigated only in *Stylopoma*, because species of *Metrarabdotos* are either extinct or rare and thus difficult to collect. Morphospecies of *Stylopoma* show excellent correspondence with genetic differences determined by protein electrophoresis (Jackson and Cheetham 1990, 1994). In addition, morphologic and genetic distances between species pairs are highly significantly correlated ($r = 0.74$, $P < 0.001$, $N = 21$; Jackson and Cheetham 1994). Thus, use of morphologic differences to distinguish species and construct hypotheses of phylogenetic relationship is well justified for *Stylopoma*. The same is assumed to apply to *Metrarabdotos*, in which morphospecies were estab-

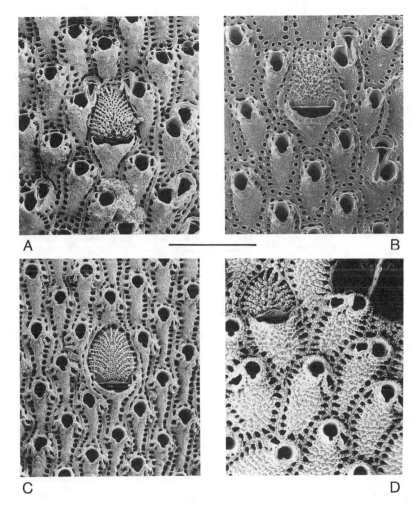

Figure 6.1   Zooidal morphology of *Metrarabdotos*, contrasting small differences between species of *tenue* group—*M.* new species 10 (A) and *M. tenue* (B)—with larger differences between species of *unguiculatum* group—*M. lacrymosum* (C) and *M. unguiculatum* (D). Scale bar 1 mm.

lished by the same procedures. In this regard, it is important to note that agreement between genetics and morphology in *Stylopoma* is maximized by splitting morphospecies to the limits of statistical significance, even though the percentage of colonies classified with high confidence declines. A "lumped" taxonomy of fewer (14), more distinct morphospecies in-

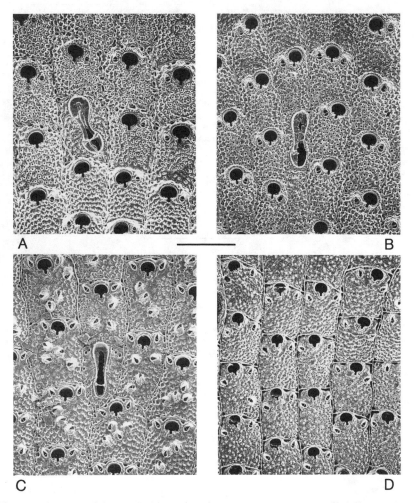

Figure 6.2   Zooidal morphology of *Stylopoma*, contrasting small differences between species of *spongites* group—*S.* new species 3 (A) and *S. spongites* (B)— with larger differences (C and D) between species of the group consisting of new species 11–15. (*S.* new species 14 is shown in C; *S.* new species 11 is shown in D.) Scale bar 0.5 mm.

cludes abundant genetic evidence for cryptic species and lower (although not statistically significantly different) correlation between genetic and morphologic distances (Jackson and Cheetham 1994).

Reconstruction of phylogenies for both *Metrarabdotos* and *Stylopoma* is complicated by uneven stratigraphic and geographic distribution of

Table 6.1

Morphospecies of Tropical American Metrarabdotos *and* Stylopoma

Taxon	Species (number of)	Characters (number of)	Morphologic Distance (mean)
Metrarabdotos	19	46	39.6
tenue group	5	46	18.8
unguiculatum group	6	46	41.4
Stylopoma	19	12	6.6
spongites group	4	12	3.7
new species 11–15	5	12	6.0

NOTE: Morphospecies based on skeletal characters; mean morphologic distance between species, square root of Malanobois D^2; species groups as in figures 6.3 and 6.4.

sampling (table 6.2). More than half the Neogene and Quaternary fossil occurrences of both genera are in an interval of detailed sampling from about 4 to 8 Ma in the Dominican Republic (figures 6.3 and 6.4), and six species of *Metrarabdotos* (new species 3 to 8, figure 6.3) occur only there. Only 10% of the occurrences of *Metrarabdotos* and 5% of *Stylopoma* are prior to 8 Ma. As pointed out by Marshall (this volume), confidence intervals on stratigraphic ranges of nine *Metrarabdotos* species appearing between 7 and 8 Ma could place their first occurrences in that earlier, less sampled interval (figure 6.3), calling into question the significance of the sequence and the apparent abruptness of their appearances in the record. The stratophenetic, nearest morphologic/stratigraphic neighbor approach thus might be less appropriate than cladistics for reconstructing *Metrarabdotos* phylogeny (Marshall, this volume). However, for the majority of species, which have two or more occurrences in the interval of detailed sampling, 95% confidence limits (calculated by the method advocated by Marshall [1990, 1991, and this volume]) extend an average of less than 0.5 My into the earlier, less sampled interval; only one extends into this interval as much as 1 My (figure 6.3).

Thus, uncertainty in the timing of species origins has little effect on the hypothesis of relationship constructed for *Metrarabdotos* by the stratophenetic approach (figure 6.3A). Species most similar in overall morphology (i.e., separated by the smallest Mahalanobis distances based on all 46 characters) and closest in stratigraphic occurrence remain viable ancestor-descendant pairs (Cheetham 1986; Cheetham and Hayek 1988; Jackson and Cheetham 1994). In the stratophenetic tree, the five species

Table 6.2

Distribution of Tropical American Fossil Metrarabdotos and Stylopoma

Interval (Ma)	Zones	Dominican Republic	Haiti	Jamaica	Trinidad	Panama	Costa Rica	Florida
0.0–2.0	NN19–21	2				3	3 (1)	(3)
2.0–3.4	NN16–18	3 (1)		1 (1)		3 (2)	6 (2)	1 (2)
3.4–4.4	NN13–15	19 (2)				(2)	1	
4.4–5.6	NN12	60 (30)				3 (4)		
5.6–8.4	NN11					7		
8.4–10.8	NN8–10				1			
10.8–13.0	NN7							
13.9–14.4	NN6	6	1					
14.4–16.2	NN5	6 (1)						
16.2–17.4	NN4							1 (1)

NOTE: Numbers listed for each area are numbers of sampled stratigraphic horizons in which *Metrarabdotos* (153 total) or *Stylopoma* (in parentheses; 51 total) occur. Panama and Costa Rica include both Caribbean and Pacific localities (Coates et al. 1992).

of the *tenue* group are separated morphologically by distances about half those for the genus as a whole (table 6.1), and they have stepped first occurrences (figure 6.3A) consistent with a close phylogenetic relationship. Relationships among the six species of the *unguiculatum* group are less certain, chiefly because of larger morphologic differences between species (table 6.1). Stratigraphic uncertainty is greatest for the comparatively uncommon group of species (new species 5 to 8) that are morphologically similar to the *tenue* group, but all of which could have originated before the interval of detailed sampling. However, the likelihood that any of the species in the interval of detailed sampling in the Dominican Republic originated elsewhere is greatly diminished by the fact that the species occurring earliest in this interval (*M. colligatum, M. auriculatum, M. lacrymosum*) are most similar morphologically to either new species 1 or new species 2, both of which occur in the Dominican Republic at about 14 Ma (figure 6.3).

Cladistic hypotheses of relationship were derived for *Metrarabdotos* using 33 of the original 46 characters with statistically significant among-species differences (Duncan's multiple-range test; Jackson and Cheetham 1994). Alternative trees were rooted either on the oldest included species (*M. micropora*) or on an appropriate living outgroup (*Escharoides costifer*, which belongs to a closely related ascophoran genus; Cheetham 1968; Jackson and Cheetham 1994). Even the shortest, most internally consistent, and least temporally discordant of the cladistic trees (figure 6.3B) is quite different from that based on stratophenetics (figure 6.3A). Extension of species ranges by 1 My or less barely begins to bridge the gap of 8 to 10 My implied by the cladogram. All four species with 95% confidence limits on first occurrences lying within the interval of detailed sampling (new species 4, 5, 9, and 10) are required to have originated at least 8 My earlier. Moreover, of the three species that do occur in this gap, only new species 2 appears closely related cladistically to any of the species in the detailed sampling interval. The morphologic and geographic proximity of new species 1 to the *tenue* group is ignored.

Thus, uncertainties in the pattern of occurrence of *Metrarabdotos* species are small compared to the differences in pattern obtained with stratophenetic and cladistic methods. Aside from disagreement with the temporal record, the cladogram also implies that the most closely related *Metrarabdotos* species are not necessarily those closest in overall morphology. However, these differences do not affect the earlier conclusion, based on the stratophenetic phylogeny, that within-species morphologic change (fluctuating stasis) is insufficient to explain the morphologic differences between species (Cheetham 1986, 1987). Indeed, the greater differences in overall morphology between putative ancestor-descendant pairs suggested by the cladistic hypothesis (figure 6.3B) would only emphasize the disparity between within- and between-species rates of change. Thus, the pattern of stasis within species relative to differences between putative ancestor-descendant species pairs, together with the temporal overlap over significant parts of their stratigraphic ranges (persistence criterion of Levinton [1988]), continues to provide strong evidence for the concentration of phenotypic evolution in *Metrarabdotos* within brief periods of cladogenesis (Cheetham 1986, 1987). The punctuated pattern of *Metrarabdotos* thus appears robust enough to form the basis for tests of speciation mode.

The stratophenetic approach is less feasible for reconstructing the phylogeny of *Stylopoma* because almost all fossil species survive to the present and all have fewer occurrences than *Metrarabdotos* (figure 6.4, table 6.2). As a result, 95% confidence limits for first occurrences average twice those for *Metrarabdotos* species. However, a strictly cladistic approach (based on characters coded as for *Metrarabdotos* [Jackson and Cheetham 1994]) also yields a stratigraphically unlikely phylogeny (figure 6.4B), with half the species required to extend back more than 17 Ma, despite the absence of *Stylopoma* in older regionally reported bryozoan-rich deposits (e.g., Canu and Bassler 1920, 1923). If two fossil occurrences classified morphologically as new species 6 and 9 are assumed to be convergent, rather than conspecific with later populations, the cladogram can be modified to agree much more closely with the 95% confidence limits on stratigraphic ranges (figure 6.4A; Jackson and Cheetham 1994).

Despite its less abundant fossil record, *Stylopoma* also shows strong

---

Figure 6.3   Stratophenetic (A) and cladistic (B) reconstructions of *Metrarabdotos* phylogeny. Stratophenetic tree is based on discriminant analysis of 46 morphologic characters (Cheetham 1986; Cheetham and Hayek 1988). Cladistic tree is the consensus of four equally most parsimonious trees based on 33 morphologic characters and rooted on *M. micropora* (Jackson and Cheetham 1994). Error bars on stratigraphic ranges are 95% confidence intervals for first occurrences of species within interval of detailed sampling (DSI) in the Dominican Republic sections (Saunders, Jung, and Biju-Duval 1986).

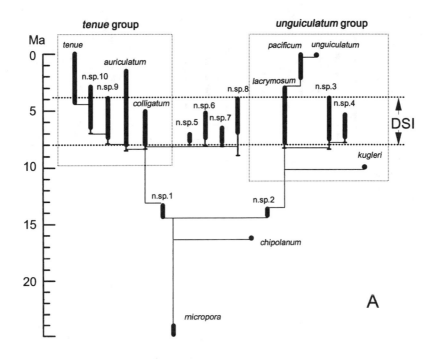

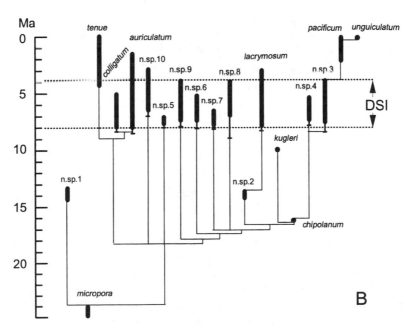

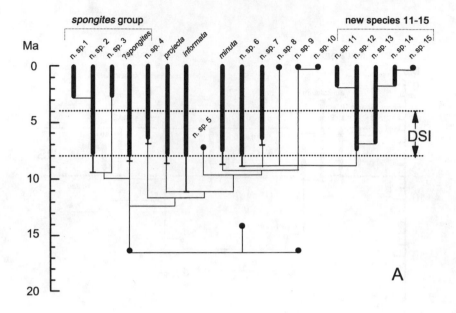

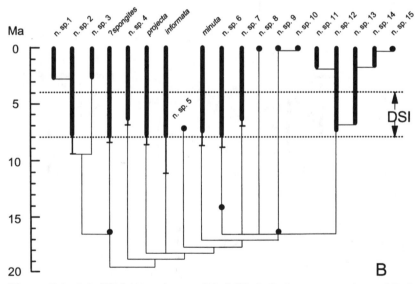

Figure 6.4 Modified (A) and unmodified (B) cladistic reconstructions of *Stylopoma* phylogeny. Both trees are the consensus of two equally most parsimonious trees based on 13 morphologic characters and rooted on *S. spongites* (Jackson and Cheetham 1994). Modified tree assumes that apparent occurrences of new species 6 and 9 at 14 to 17 Ma are morphologically convergent, rather than conspecific with younger populations of those species.

evidence for a punctuated pattern of speciation. Within-species stasis is indicated by the fact that the great majority of fossil and living occurrences, including the first occurrences of 11 of the 19 species, are referred to their respective morphospecies with $P > 0.9$ (Jackson and Cheetham 1994). In addition, temporal overlap between putative ancestor-descendant species pairs is even greater than for *Metrarabdotos*, with ten species surviving beyond the detailed sampling interval more than 6 My to the Holocene. Consistency between the two genera suggests that punctuated patterns could be characteristic of cheilostomes in general, and *Stylopoma* is included in the tests of speciation mode despite its less abundant fossil record.

Rate Test for Selection Versus Random Change

The test employed here is based on the neutral model of phenotypic evolution, that is, the null hypothesis that observed morphologic differences between species result from mutation and genetic drift as the sole evolutionary forces (Lynch 1990). The test statistic summarizes the rate of evolution over the entire time of divergence; thus, for episodic patterns like those of *Metrarabdotos* and *Stylopoma,* the results require careful interpretation.

The method is based on the quantitative genetic assumption that traits are polygenic, differing by small effects at many loci. It partitions phenotypic variance in each log-transformed trait into between species (var_B) and within-species (var_W) components, obtained by standard one-way ANOVA. The rate statistic

$$\Delta = (var_B/var_W)/t$$

is the ratio of variance components divided by the number of generations, *t,* over which species diverged (Lynch 1990, equation 3). This statistic is tested against the null expectation of the neutral model, namely that the rate of phenotypic change is equal to the rate of accumulation of nondeleterious mutations, measured by the rate of mutational input to within-species phenotypic variance.

To use this test, Δ must be compared to typical rates of mutational input to phenotypic variance. Lynch (1988) compiled rates for a wide variety of organisms, including invertebrates, for traits assumed to be controlled by polygenic inheritance. He found that rates of input typically lie over two orders of magnitude, between 10^{-2} and 10^{-4} per generation. Thus, values of Δ greater than 10^{-2} imply directional selection and values less than 10^{-4} imply stabilizing selection.

To convert divergence times to t, the number of generations, we used one- and ten-year generation times. In breeding experiments, the rapidly growing encrusting colonies of *Stylopoma* reach reproductive maturity in about one year. By analogy with other, more slowly growing erect chei-lostomes (Stebbing 1971), *Metrarabdotos* generation times could be as long as ten years.

We followed Lynch (1990) in averaging values of Δ across traits to obtain mean evolutionary rates in each genus. This procedure is justified because calculations based on a null hypothesis of random change are less dependent on correlations among traits than those based on selection (Lande 1979). However, comparison between values of Δ for individual species pairs averaged across traits with those based on canonical scores from pairwise discriminant analysis suggests that trait covariances may not always have negligible effects (Cheetham, Jackson, and Hayek 1994). Canonical scores are not an option here because multiple-species discrim-inant analysis does not reduce to a single canonical variate.

The rates of morphologic divergence obtained for both *Metrarabdotos* and *Stylopoma* and their constituent species groups (table 6.3) have a wide range of values across traits, yielding standard deviations that equal or exceed means, as has been found for the mammalian groups analyzed by Lynch (1990). However, mean values of Δ are all well below 10^{-4}, the minimum rate consistent with the neutral model (table 6.3). Thus, sta-bilizing selection is required to explain the mean rate of divergence for each genus as a whole and its constituent species groups. Generation times shorter than ten years for *Metrarabdotos* or one year for *Stylopoma*, or attenuation of divergences over longer time intervals in either genus (e.g., as required by the cladistic hypotheses in figures 6.3B and 6.4B), would only emphasize these results. Despite the variation in rates for individual traits, all 12 characters in *Stylopoma* are consistent with these results. In *Metrarabdotos*, for which standard deviations are higher than for *Stylo-poma* (table 6.3), the rates for all 46 individual traits in the genus as a whole are also consistently under 10^{-4}. However, in each of the two species groups, one character (number of oral denticles in the *tenue* group, length of avicularium on zooids at bifurcation of rows in the *unguiculatum* group [Cheetham 1986, 1987]), evolved at a rate exceeding the min-

Table 6.3

Divergence Times and Rates in Metrarabdotos and Stylopoma

Taxon	Minimum Divergence Time (Ma)	Maximum Rate of Divergence (Δ)		
		Mean	S. D.	Range
METRARABDOTOS	16.5	6.0×10^{-6}	1.2×10^{-5}	2.5×10^{-7} to 6.0×10^{-5}
tenue group	3.5	9.7×10^{-5}	3.3×10^{-5}	6.9×10^{-8} to 2.1×10^{-4}
unguiculatum group	10.0	3.5×10^{-5}	1.2×10^{-4}	1.3×10^{-7} to 6.4×10^{-4}
STYLOPOMA	16.5	1.1×10^{-7}	8.2×10^{-8}	2.3×10^{-8} to 3.0×10^{-7}
spongites group	8.0	1.1×10^{-7}	1.2×10^{-7}	3.1×10^{-9} to 3.7×10^{-7}
new species 11–15	7.0	4.8×10^{-7}	4.8×10^{-7}	4.4×10^{-8} to 1.5×10^{-6}

NOTE: Maximum rates of divergence are averaged across 46 characters in *Metrarabdotos* and 12 in *Stylopoma*. Assumed generation times are 1 year for *Stylopoma*, 10 years for *Metrarabdotos*.

imum neutral expectation, but well short of the maximum (i.e., not requiring directional selection; table 6.3).

Temporal Scaling and Initial Rates of Divergence

Mean rates of morphologic divergence in *Metrarabdotos,* irrespective of whether based on one- or ten-year generation times, all fall within the range of values obtained by Lynch (1990) for skeletal characters in a wide variety of mammalian groups (figure 6.5). Those for *Stylopoma* are all lower than the slowest *Metrarabdotos* and mammalian rates, despite the fact that the rate of origination for *Stylopoma* species is approximately the same as for *Metrarabdotos* over much of the Neogene and *Stylopoma* maintained greater species richness through the late Neogene and Quaternary because of its much lower extinction rate (figures 6.3 and 6.4; Cheetham and Jackson 1995). The difference in Δ results from the greater morphologic distances between species in *Metrarabdotos* (table 6.1).

The mammalian data used by Lynch (1990) to calculate Δ include a broad range of time scales over which each group diverged, from 10^2 to 10^3 generations for *Homo sapiens* to 10^7 to 10^8 generations for squirrels. Lynch found that differences in mammalian rates can be explained by a strong log-linear relation between the total amount of morphologic divergence within a group and the duration of the interval over which it diverged, implying a decline in rate over time. Thus, he inferred that the high values of Δ for *Homo sapiens,* rather than being unusual, may also have characterized the first few hundred generations in the divergence of species in all mammalian groups.

For *Metrarabdotos,* differences in Δ between the *tenue* and *unguiculatum* groups and the genus as a whole also appear to reflect the temporal scaling observed by Lynch. With ten-year generation times, total divergence values fall near the log-linear regression line through the mammalian data (figure 6.6A). For generation times closer to one year, *Metrarabdotos* values scale to a somewhat lower rate than for mammals. However, in either case, initial rates of divergence of species occurring within the first 10^3 generations would be closer to those for *Homo sapiens.* This time scale is consistent with the abrupt appearances of *Metrarabdotos* species in the fossil record.

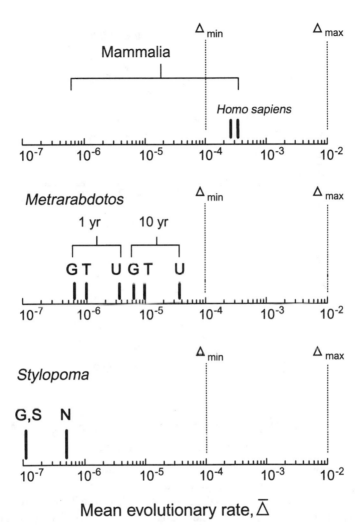

Figure 6.5 Estimates of evolutionary rates in *Metrarabdotos* and *Stylopoma* compared with those for 12 mammalian groups, including *Homo sapiens* (Lynch 1990). Estimates for *Metrarabdotos* (G, whole genus; T, *tenue* group; U, *unguiculatum* group) are based on one-year and ten-year generation times; those for *Stylopoma* (G, whole genus; N, new species 11–15; S, *spongites* group) are all based on one-year generation times.

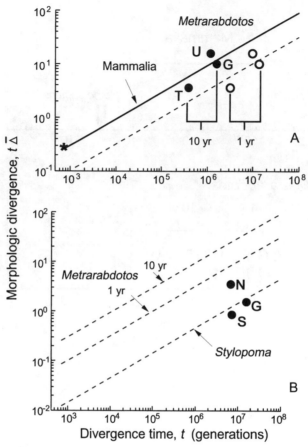

Figure 6.6 Relationship between amount and duration of morphologic divergence. Graph A shows the apparent fit between *Metrarabdotos* divergence (ten-year generation times are filled circles; G, whole genus; T, *tenue* group; U, *unguiculatum* group) and regression based on mammalian data (solid line; $t\Delta = 0.0067t^{0.52}$; *, *Homo sapiens*; Lynch 1990). Divergence in *Metrarabdotos* based on one-year generation times (open circles) may fit lower (dashed) line drawn with same slope (0.52), but lower intercept (0.0017). Graph B shows the possible fit of *Stylopoma* divergence (G, whole genus; N, new species 11–15; S, *spongites* group) to a still lower line with slope 0.52 and intercept 0.00032.

Temporal scaling of divergence rates is less obvious for *Stylopoma* because divergence times of species groups are less variable (figure 6.6B, table 6.3). However, in keeping with the smaller morphologic differences between species (table 6.1), Δ is clearly lower than for *Metrarabdotos* and mammals at equivalent divergence times, including the first 10^3 generations (figure 6.6B).

The most rapid rates of divergence that can be projected for species of either *Metrarabdotos* or *Stylopoma* (assuming the same slope for the mammalian log-linear regression and the hypothesized bryozoan lines in figure 6.6) would fall within the range of expectation of the neutral model (figure 6.7). Divergence times shorter than a few thousand generations for *Metrarabdotos* and a few tens of generations for *Stylopoma* all yield values of Δ between 10^{-4} and 10^{-2}. The implication is that mutation and random drift alone are sufficient to account for the morphologic differences between species in both the *Metrarabdotos* and the *Stylopoma* clades, with no need to invoke directional selection.

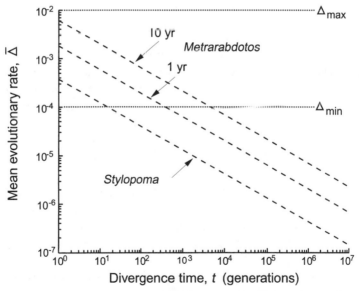

Figure 6.7 Hypothesized decline in rates of morphologic divergence in *Metrarabdotos* and *Stylopoma* based on divergence/time regressions in figure 6.6.

Implications and Limitations for Mode of Speciation

Strong evidence for the prevalence of stabilizing selection in the long-term morphologic evolution of *Metrarabdotos* and *Stylopoma* confirms their punctuated patterns of speciation, regardless of the precision with which the times and places of species origin are preserved in the record. Likewise, the close fit of rates of divergence in *Metrarabdotos* to the temporal scaling of rates in mammals obtained by Lynch (1990) and the apparent scaling of rates in *Stylopoma* to an even slower pace provide considerable assurance that morphologic change in these bryozoans was indeed confined to geologically brief cladogenetic events within the limits of stratigraphic resolution. This evidence, together with the genetic evidence associating morphospecies with genetic differences in *Stylopoma,* is thus consistent with the basic tenet of punctuated equilibria theory that most evolutionary change is concentrated in cladogenetic events, rather than accumulating over major intervals of a species' duration (Eldredge and Gould 1972).

Lynch (1990, p. 737) proposed the following model to explain the similar results he obtained for mammals:

> Consider the simple situation in which reproductively isolated populations remain subject to the same Gaussian stabilizing selective pressures, leading to an optimum phenotype . During the initial phases of isolation, sampling error resulting from finite population size causes the population means to drift apart. Provided that the mean phenotype of the ancestral population was near the optimum, the initial divergence would be relatively rapid because of the flatness of the fitness function in the neighborhood of the optimum. However, as the means drift farther from the optimum, the selection gradient becomes steeper and the rate of divergence must decline. Eventually, a balance is struck between the diversifying forces of drift and mutation and the stabilizing forces of selection, such that the expected between-population variance no longer increases.

The evidence for stabilizing selection as the source of long-term within-species stasis in *Metrarabdotos* and *Stylopoma* is strong because it is based on rejection of the null hypothesis of neutral expectation. Acceptance of this null hypothesis for the short-term origin of between-species differences is weaker evidence for the sufficiency of random genetic change, even though values of $\Delta > 10^{-2}$ cannot be projected for even the shortest time scales over which ancestor-descendant species pairs could have diverged in either genus. More rigorous evaluation of the role of directional selection in the diversification of these genera is contingent upon estimating the time scales at which unrealistically high levels of selective mortality are needed to explain the observed amounts of morphologic divergence. Calculation of selection intensities requires simplifying assumptions concerning the mode of operation of natural selection (i.e., truncation) and estimates of genetic parameters (i.e., heritabilities and genetic covariances) (Lande 1979). Although truncation is not reasonable as a model of natural selection, it provides an estimate of the minimum intensity required to produce a given phenotypic change over a given number of generations, and thus is the best available method for evaluating the role of directional selection (Manly 1985). Applied to mammalian breeding data, this method has yielded values indicating the feasibility of directional selection to produce observed phenotypic differences between species on time scales as short as 10^3 generations (Lofsvold 1988).

The problem of obtaining estimates of genetic parameters is not one that can be simply solved by substituting phenotypic values into calculations of selection intensity (Willis, Coyne, and Kirkpatrick 1991). Even though evidence from several groups of animals suggests that phenotypic variances and covariances can be similar to genetic ones, the relationship is at best one of proportionality not equality (Lofsvold 1986; Cheverud 1988). Thus, estimates of trait heritabilities at least are required. Partitioning variance in zooid morphology into within- and between-colony components has been thought to offer a basis for estimating heritability in Bryozoa (Pachut 1989; see also Anstey and Pachut, this volume), but this relationship has only recently been studied empirically with breeding data (Cheetham, Jackson, and Hayek 1993). Estimation of these parameters in living species of *Stylopoma* and application of the inferred relationship to fossil species of *Metrarabdotos* (Cheetham, Jackson, and Hayek 1994) suggest that heritabilities and genetic covariances for traits of skeletal morphology in these genera are typical of polygenic traits in other taxa (Cheverud 1988), and distinctly smaller than for quantitative traits that have been shown to differ by single shifts in a major gene (Smith 1993). At the very least, the results obtained with bryozoan data indicate that the application of quantitative genetic methods is a valid approach to the inference of the speciation process from phenotypic patterns observed in the fossil record.

Summary: Pattern and Process in Speciation

Close correspondence between morphospecies and genetic differences in living *Stylopoma* makes it highly likely that punctuated patterns of morphologic change preserved in the Neogene and Quaternary records of *Stylopoma* and *Metrarabdotos* in tropical America reflect evolution at the species level. Uncertainties in timing and geographic location of species origins appear insufficient to contradict the geologically abrupt appearances of species in the record, the static morphologies of species for millions of years, and the extensive temporal overlap between apparent ancestor-descendant species pairs.

Tests for evolutionary mechanisms underlying these patterns, based on quantitative genetic theory, reveal that rates of phenotypic evolution in *Metrarabdotos* and *Stylopoma* are equal to or slightly below those for mammals at equivalent divergence times. Long-term evolutionary rates in these cheilostomes are too slow to have resulted from random genetic change (mutation and drift) alone, implying prevalence of stabilizing selection reflected in the morphologic stasis within species. Thus, evolutionary change must be concentrated within brief episodes of cladogenesis, consistent with the abrupt appearances of species in the record and with the basic tenet of punctuated equilibria theory associating evolutionary change with speciation.

If the relationship between the amount and duration of divergence observed for mammals applies to these cheilostomes, their short-term evolutionary rates appear consistent with the hypothesis that morphologic differentiation of species resulted entirely from mutation and random genetic drift, without input from directional selection. However, unlike the conclusion regarding prevalence of stabilizing selection in the long term, this explanation for rapid short-term change is only a first approximation based on acceptance of the null hypothesis of sufficiency of random genetic change. The possible role of directional selection in the punctuated patterns of these cheilostomes can be evaluated more rigorously now that empirical evidence from breeding experiments supports a means of estimating genetic parameters of fossils.

Acknowledgments   We thank JoAnn Sanner for morphologic measurements and other help; Yira Ventocilla for processing samples; Peter Jung, John Saunders, Emily Vokes, Kevin Schindler, and Laurel Collins for material and stratigraphic documentation; Amalia Herrera, Javier Jara, David West, Steve Miller, Julio Calderon, and Marcos Sorriano for help with Recent specimens; Lee-Ann Hayek for statistical advice; Scott Lidgard and Doug Erwin for helpful comments on the manuscript; and the Kuna nation and Government of Panama for permitting work in the San Blas. This work was supported by the Smithsonian Scholarly Studies Program, the Marie Bohrn Abbott Fund of the National Museum of Natural History, and the Smithsonian Tropical Research Institute.

References

Barton, N. H. and B. Charlesworth. 1984. Genetic revolutions, founder effects, and speciation. *Annual Review of Ecology and Systematics* 15:133–164.

Canu, F. and R. S. Bassler. 1920. North American early Tertiary Bryozoa. *United States National Museum Bulletin* 106:1–879.

———. 1923. North American later Tertiary and Quaternary Bryozoa. *United States National Museum Bulletin* 125:1–302.

Cheetham, A. H. 1968. Morphology and systematics of the bryozoan genus *Metrarabdotos*. *Smithsonian Miscellaneous Collections* 153(1):1–121.

Cheetham, A. H. 1986. Tempo of evolution in a Neogene bryozoan: Rates of morphologic change within and across species boundaries. *Paleobiology* 12:190–202.

———. 1987. Tempo of evolution in a Neogene bryozoan: Are trends in single morphologic characters misleading? *Paleobiology* 13:286–296.

Cheetham, A. H. and L. C. Hayek. 1988. Phylogeny reconstruction in the Neogene bryozoan *Metrarabdotos*: A paleontological evaluation of methodology. *Historical Biology* 1:65–83.

Cheetham, A. H. and J. B. C. Jackson. 1995 in press. Speciation, extinction, and the decline of erect growth in Neogene and Quaternary cheilostome Bryozoa of tropical America. In J. B. C. Jackson, A. G. Coates, and A. F. Budd, eds., *Evolution and Environment in Tropical America*. Chicago: University of Chicago Press.

Cheetham, A. H., J. B. C. Jackson, and L. C. Hayek. 1993. Quantitative genetics of bryozoan phenotypic evolution. I. Rate tests for random change versus selection in differentiation of living species. *Evolution* 47:1526–1538.

———. 1994. Quantitative genetics of bryozoan phenotypic evolution. II. Analysis of selection and random change in fossil species using reconstructed genetic parameters. *Evolution* 48:360–375.

Cheverud, J. M. 1988. A comparison of genetic and phenotypic correlations. *Evolution* 42:958–968.

Coates, A. G., J. B. C. Jackson, L. S. Collins, T. M. Cronin, H. J. Dowsett, L. M. Bybell, P. Jung, and J. A. Obando. 1992. Closure of the Isthmus of Panama: The near-shore marine record of Costa Rica and western Panama. *Geological Society of America Bulletin* 104:814–828.

Eldredge, N. and S. J. Gould. 1972. Punctuated equilibria: An alternative to phyletic gradualism. In T. J. M. Schopf, ed., *Models in Paleobiology*. San Francisco: Freeman, Cooper.

Geary, D. H. 1991. Patterns of evolutionary tempo and mode in the radiation of *Melanopsis* (Gastropoda; Melanopsidae). *Paleobiology* 16:492–511.

———. 1992. An unusual pattern of divergence between two fossil gastropods: Ecophenotypy, dimorphism, or hybridization? *Paleobiology* 18:93–109.

Gould, S. J. 1991. Opus 200. *Natural History* 8/91:12–18.

Jackson, J. B. C. and A. H. Cheetham. 1990. Evolutionary significance of morphospecies: A test with cheilostome Bryozoa. *Science* 248:579–583.

———. 1994. Phylogeny reconstruction and the tempo of speciation in cheilostome Bryozoa. *Paleobiology* 20:407–423.

Kat, P. W. and G. M. Davis. 1983. Speciation of molluscs from Turkana Basin. *Nature* 304:660–661.

Lande, R. 1979. Quantitative genetic analysis of multivariate evolution, applied to brain:body size allometry. *Evolution* 33:402–416.

———. 1985. Expected time for random genetic drift of a population between stable phenotypic states. *Proceedings of the National Academy of Sciences of the U.S.A.* 82:7641–7645.

———. 1987. The dynamics of peak shifts and the pattern of morphological evolution. *Paleobiology* 12:343–354.

Larson, A. 1989. The relationship between speciation and morphological evolution. In D. Otte and J. A. Endler, eds., *Speciation and its Consequences*. Sunderland, Mass.: Sinauer.

Levinton, J. 1988. *Genetics, Paleontology and Macroevolution*. Cambridge: Cambridge University Press.

Lofsvold, D. 1986. Quantitative genetics of morphological differentiation in *Peromyscus*. I. Tests of homogeneity of genetic covariance structure among species and subspecies. *Evolution* 40:559–573.

———. 1988. Quantitative genetics of morphological differentiation in *Peromyscus*. II. Analysis of selection and drift. *Evolution* 42:54–67.

Lynch, M. 1988. The rate of polygenic mutation. *Genetical Research Cambridge* 51:137–148.

———. 1990. The rate of morphological evolution in mammals from the standpoint of the neutral expectation. *American Naturalist* 136:727–741.

MacLeod, N. 1991. Punctuated anagenesis and the importance of stratigraphy to paleobiology. *Paleobiology* 17:167–188.

Malmgren, B. A. and J. P. Kennett. 1981. Phyletic gradualism in a Late Cenozoic planktonic foraminiferal lineage; DSDP Site 284, southwest Pacific. *Paleobiology* 7:230–240.

Manly, B. F. J. 1985. *The Statistics of Natural Selection of Animal Populations.* London and New York: Chapman and Hall.

Marshall, C. R. 1990. Confidence intervals on stratigraphic ranges. *Paleobiology* 16:1–10.

———. 1991. Estimation of taxonomic ranges from the fossil record. In N. L. Gilinsky and P. L. Signor, eds., *Analytical Paleobiology.* Knoxville, Tennessee: The Paleontological Society.

Otte, D. and J. A. Endler, eds. 1989. *Speciation and Its Consequences.* Sunderland, Mass.: Sinauer.

Pachut, J. F. 1989. Heritability and intraspecific heterochrony in Ordovician bryozoans from environments differing in diversity. *Journal of Paleontology* 63:182–194.

Palmer, A. R. 1985. Quantum changes in gastropod shell morphology need not reflect speciation. *Evolution* 39:699–705.

Reyment, R. A. 1985. Phenotypic evolution in a lineage of the Eocene ostracod *Echinocythereis. Paleobiology* 11:174–194.

Saunders, J., P. Jung, and B. Biju-Duval. 1986. Neogene paleontology in the northern Dominican Republic. 1. Field surveys, lithology, and age. *Bulletins of American Paleontology* 89:1–79.

Smith, T. B. 1993. Disruptive selection and the genetic basis of bill size polymorphism in the African finch *Pyrenestes. Nature* 363:618–620.

Stebbing, A. R. D. 1971. Growth of *Flustra foliacea* (Bryozoa). *Marine Biology* 9:267–272.

Turner, J. R. G. 1986. The genetics of adaptive radiation: A neo-Darwinian theory of punctuational evolution. In D. M. Raup and D. Jablonski, eds., *Patterns and Processes in the History of Life.* Berlin and Heidelberg: Springer-Verlag.

Wei, K.-Y. and J. P. Kennett. 1988. Phyletic gradualism and punctuated equilibrium in the late Neogene planktonic foraminiferal clade *Globoconella. Paleobiology* 14:345–363.

Willis, J. H., J. A. Coyne, and M. Kirkpatrick. 1991. Can one predict the evolution of quantitative characters without genetics? *Evolution* 45:441–444.

Wright, S. 1982. Character change, speciation, and the higher taxa. *Evolution* 36:427–443.

7

Stratigraphy, the True Order of Species Originations and Extinctions, and Testing Ancestor-Descendant Hypotheses Among Caribbean Neogene Bryozoans

Charles R. Marshall

The fossil record provides the only direct evidence for the order of appearances and disappearances of species. However, because that record is incomplete, the observed order of first and last appearances may not accurately reflect the true order of originations and extinctions. Species that appear sequentially in the fossil record may have actually shared the same time of origin, or even originated in the reverse order. Similarly, species that disappear successively in the fossil record may have become extinct simultaneously, or in the opposite order. The degree to which patterns of first and last appearances reflect the true order of originations and extinctions depends on many factors, particularly on the richness of the fossil record and the size of the stratigraphic gap separating the times of first (or last) appearances of the species being compared. Thus, there would be little dispute that trilobites really originated and became extinct before the origin of dinosaurs. But in cases where first appearances, or disappearances, occur within a few million years of one another the reliability of the fossil record is not so clear.

Here I describe a method for calculating the probability that, for a species pair, the observed sequence of appearances or disappearances accurately reflects the true order of origination or extinction. The observed stratigraphic range and the numbers of fossil horizons are used as a measure of the richness of the fossil record, and in light of this richness the significance of the mismatch in the times of first or last appearance is evaluated.

This analytic approach is used to test Cheetham's (1986) stratigraphically based hypotheses of ancestor-descendant relationship among the Neogene bryozoan species of the genus *Metrarabdotos*. A striking aspect of the sedimentary record where *Metrarabdotos* fossils are found is a virtual absence of rock (unconformities in the majority of sections) for approximately 6 My preceding the largest radiation of species at 8 Ma (Saunders, Jung, and Biju-Duval 1986). It is conceivable that most of these species were already extant prior to 8 Ma, but are undetectable given the absence of rock of the right age. There may have been no radiation, and the observed order of appearance in the fossil record might carry no phylogenetic information but may simply be an effect of incomplete preservation and the hiatus. I tested this hypothesis, using the quantitative approach developed by Springer (1990).

I thus used two methods to estimate the value of stratigraphic data for inferring ancestor-descendant relationships among the species of *Metrarabdotos*. However, it should be noted that the major conclusions of Cheetham's (1986) classic study, the demonstration of stasis within and punctuated change between species, are not drawn into question by this study.

When developing quantitative descriptions of natural phenomena, there is often a tension between the limitations of mathematical description and tractability on one hand, and a realistic or comprehensive characterization of the phenomena on the other. The model developed here can be used safely only in those instances in which the probability of finding a fossil is equal throughout its true stratigraphic range. This requirement will not, however, be universally met and thus must be tested before one applies the methods described here. Notably, the requirement of uniform probability of a find is likely to be violated in cases where there are major hiatuses (unconformities) within the range, or when there are pronounced secular trends in sedimentation (e.g., in cases where a taxon's ranges lies in a shallowing up sequence), or when there are pronounced secular changes in species abundances with time, etc.

Materials and Methods

Metrarabdotos Fossil Record

The primary purpose of Cheetham (1986) was an examination of the nature of morphologic change in fossils of the bryozoan *Metrarabdotos* collected in the Dominican Republic. Hence that study was restricted to specimens of sufficient quality for morphological analysis, and fossils with poorer preservation were excluded. In my own work, all *Metrarabdotos* fossil horizons—not just those yielding the better preserved specimens—are analyzed. The absolute ages of the strata yielding *Metrarabdotos* fossils are given in appendix A.

The stratigraphy of the Dominican Republic is well understood (Saunders et al. 1986). Of particular importance is the virtual absence of rock during the interval 8 to about 14 million years ago. The rock record is relatively uniformly complete from 4–8 Ma. (At a temporal resolution of 0.16 My, the mean interval between successively sampled populations, the expected completeness is 0.63 [Cheetham 1986].). But the record begins to deteriorate less than 4 Ma, and is largely absent from 3.5 Ma to the present. Of the 18 species in Cheetham's (1986) study, 13 are known from two or more fossil horizons; of these, 11 first appear in rocks younger than 8 million years. The two older taxa known from two or more fossil horizons are found where the rock record is too patchy for the methods described here to be effectively applied. Thus my study is restricted to the 11 species known from at least two fossil horizons that first appear in rocks younger than 8 Ma.

Statistical Semantics

From a statistical standpoint the true beginning of a stratigraphic range is a fixed parameter; it lies at a fixed point in time with probability 1.0, with everywhere else having a probability of 0.0 (Strauss and Sadler 1989). Hence quantitative assessments of the incompleteness of the fossil record are based on confidence intervals, not probability distributions which are only appropriate when estimating the values of random variables. To state that the true end of a stratigraphic range lies between the base of the known range and the end of the 95% confidence interval with a probability 0.95 is incorrect, since the true end-point either does ($p = 1.0$) or does not ($p = 0.0$) lie within the confidence interval. However, it is correct to say that the probability is 0.95 that the true end-point of the range lies somewhere within the 95% confidence interval.

In the sections following, I calculate the confidence that for two taxa, given the size of the stratigraphic gap between their times of first appearance and the richness of their fossil records, the true base of the later appearing taxon actually lies stratigraphically above the true base of the earlier appearing taxon. This mode of expression is cumbersome, however. So for the sake of clarity I often use the vocabulary of probability distributions: I talk of the probability that the stratigraphic position of the true base of the later appearing taxon is younger than the true base of the earlier appearing taxon. The reader is alerted to the fact that this convenient choice of words nevertheless incorrectly implies that the true bases of the taxa are random variables, rather than fixed parameters.

Establishing the Order of Origination from Fossil Data

Given a pair of taxa, we can calculate the probability that the observed order of appearance in the rock record reflects the true order of origination; did the later appearing taxon really originate later than the earlier appearing taxon? Thus we need to calculate the probability that the stratigraphic position of the true but unknown base (time of origin) of the later appearing taxon (B_1) is younger than the true base of the earlier appearing taxon (B_e)—that is, we need to calculate $P(B_1$ above $B_e)$. In the equations that follow, R_1, H_1, R_e, and H_e designate the observed stratigraphic ranges and numbers of fossil horizons of the later and earlier appearing taxa respectively, and m is the gap between the times of first appearance of the two taxa (figure 7.1A). Note that stratigraphic distance in these equations is measured from the base of the later appearing taxon, and that it increases down section (figure 7.1). The equation for the probability $P(B_1$ above $B_e)$ consists of two parts: Part 1 covers the possibility that the true base of the later appearing taxon lies above the time of first appearance of the earlier appearing taxon [$P(B_1$ above B_e; if $0 < B_1 < m$), figure 7.1A]. Part 2 covers the possibility that the true base of the later appearing taxon occurs below the first appearance of the earlier appearing taxon [$P(B_1$ above B_e; if $B_1 > m$), figure 7.1B]. The overall probability, $P(B_1$ above $B_e)$, is the sum of these two probabilities.

Equation for part 1 of $P(B_1$ above $B_e)$: If the later appearing taxon really originated anywhere between its first appearance and m My below its base (shaded area in figure 7.1A) then it must have originated later than the earlier appearing taxon. Confidence that the later appearing taxon really originated in this interval is given by an equation developed by

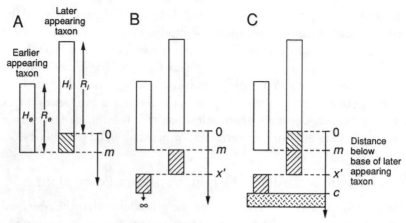

Figure 7.1   Scheme used to derive the equation for the probability that the later appearing taxon really originated later than the earlier appearing taxon. Observed stratigraphic ranges are designated by open boxes. A: If the later appearing taxon had its origin anywhere above the base of the earlier appearing taxon (hatched box) then it must have originated later than the earlier appearing taxon. B: If the later appearing taxon had its origin somewhere below the base of the earlier appearing taxon, then the later appearing taxon really was the later to originate if, for all values of x', the later appearing taxon originated above x' and the earlier appearing taxon originated below x'. C: If there is a hiatus in the fossil record c My below the base of the later appearing taxon, then the double integral in equation-3 is only evaluated to c rather than ∞.

Strauss and Sadler (1989) for calculating a confidence interval on one endpoint of a stratigraphic range:

$$P(B_l\ above\ B_e;\ if\ 0 < B_l < m) = 1 - \left(1 + \frac{m}{R_l}\right)^{-(H_l\ -\ 1)} \qquad \text{(eq. 1)}$$

Equation for part 2 of $P(B_1\ above\ B_e)$: If the later appearing taxon really originated somewhere *below* the base of the earlier appearing taxon (figure 7.1B), then confidence that later appearing taxon actually originated later than the earlier appearing taxon can be calculated with the following equation (see appendix B for a description of how I derived this equation):

$P(B_l \text{ above } B_e; \text{ if } B_l > m) =$

$$\left(\frac{H_e - 1}{R_e}\right)\left(\frac{H_l - 1}{R_l}\right) \int_m^\infty \int_m^{x'} \left(1 + \frac{x}{R_l}\right)^{-H_l} \left(1 + \left(\frac{x' - m}{R_e}\right)\right)^{-H_e} dx \, dx'$$

(eq. 2)

Thus the overall probability that the later appearing taxon originated later than the earlier appearing taxon, $P(B_l \text{ above } B_e)$, is given by the sum of equation-1 and equation-2:

$$P(B_l \text{ above } B_e) = 1 - \left(1 + \frac{m}{R_l}\right)^{-(H_l - 1)} +$$

$$\left(\frac{H_e - 1}{R_e}\right)\left(\frac{H_l - 1}{R_l}\right) \int_m^\infty \int_m^{x'} \left(1 + \frac{x}{R_l}\right)^{-H_l} \left(1 + \left(\frac{x' - m}{R_e}\right)\right)^{-H_e} dx \, dx'$$

(eq. 3)

Double Integrals Are Easily Solved

Equation-3 is not readily solved by hand. But with software packages such as *Theorist* (for Macintosh) or *Mathematica* (which will run on a wide variety of computers, including Macintosh, UNIX, and IBM systems), it can usually be solved in under a minute. *Theorist* was used for the calculations in this paper. In *Theorist* the equation is typed into the computer, using the menus for entering the integral signs, etc. The equation is solved using the "calculate" command under the "manipulate" menu, and the answer is printed to the screen shortly afterward. The code for solving equation-3 in *Mathematica*, with a worked example, is given in appendix C.

The properties of equation-3 were explored by solving the equation for three hypothetical pairs of taxa (figure 7.2). The graphs show that the equation does behave in accord with stratigraphic intuition. The salient features to note are:

- If the ranges and number of fossil horizons for both taxa in the pair are equal and if both first appear in the fossil record simultaneously, there is a 50% chance that either originated first (y intercept, figure 7.2A).

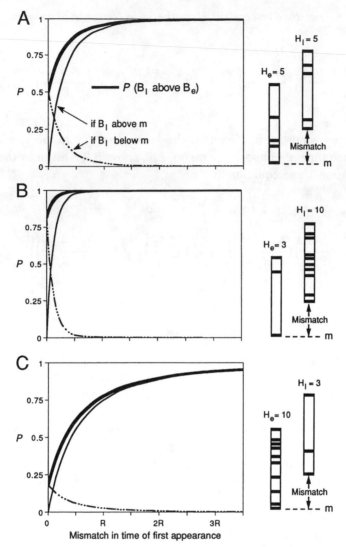

Figure 7.2 Probability that the later appearing taxon really originated later than the earlier appearing taxon (heavy line), as a function of the mismatch (*m*) between the times of first appearance of the two species (equation-3). This probability, $P(B_l$ above $B_e)$, equals the sum of two probabilities: first, the probability that the later appearing taxon really originated later than earlier appearing taxon, if the later appearing taxon appeared above the base of the earlier appearing taxon (fine complete lines, equation-1); second, the probability that the later appearing taxon really originated later than earlier appearing taxon, if the later appearing taxon appeared below the base of the earlier appearing taxon (fine broken lines, equation-2). For each of the hypothetical

- If the later appearing taxon has a richer fossil record than the earlier appearing taxon, then even with little mismatch in their times of first appearance in the fossil record, we have high confidence that the later appearing taxon really was the later to originate (figure 7.2B).
- If the later appearing taxon has a poorer fossil record than the earlier appearing taxon, then even with a sizable mismatch in the time of first appearance in the fossil record, there is a reasonable probability that the later appearing taxon actually was the first to originate (figure 7.2C).

Useful variants of equation-3 may also be derived. For example, the probability that the earlier appearing taxon really originated later than the later appearing taxon, $P(B_l$ below $B_e)$, is given by (see appendix B for a derivation):

$$P(B_l \text{ below } B_e) =$$
$$\left(\frac{H_e - 1}{R_e}\right)\left(\frac{H_l - 1}{R_l}\right) \int_m^\infty \int_{x'}^\infty \left(1 + \frac{x}{R_l}\right)^{-H_l} \left(1 + \left(\frac{x' - m}{R_e}\right)\right)^{-H_e} dx\, dx'$$

(eq. 4)

Dealing with Hiatuses in the Rock Record

In my earlier analysis of the fossil record of *Metrarabdotos* it became apparent that the hiatus in the rock record that immediately predates the largest radiation of *Metrarabdotos* species presents special problems for the application of confidence intervals (Marshall 1990a). However, equation-3 is sufficiently general that it can be adapted to deal with hiatuses in the rock record. The probability that the later appearing taxon actually originated later than the earlier appearing taxon is only meaningful when calculated in the portion of the stratigraphic column where fossils of the later appearing taxon may potentially be found. Hence, if there is an

---

examples, the stratigraphic ranges (R) of the earlier and later appearing taxa are equal, and are indicated by the open boxes. A: case where both earlier and later appearing taxa are known from 5 fossil horizons; B: case where earlier appearing taxon is known from 10 horizons, and later appearing taxon from 3 horizons; C: case where earlier appearing taxon is known from 3 horizons, and later appearing taxon from 10 horizons.

absence of rock (or the appropriate rock type) that extends for c My below the base of the latest appearing taxon (figure 7.1C), the limits of the integral in equation-2 must be modified, and equation-3 becomes (see also appendix B):

$$P(B_l \text{ above } B_e; \text{ truncation } c \text{ myr below } B_l) = 1 - \left(1 + \frac{m}{R_l}\right)^{-(H_l - 1)} +$$

$$\left(\frac{H_e - 1}{R_e}\right)\left(\frac{H_l - 1}{R_l}\right) \int_m^c \int_m^{x'} \left(1 + \frac{x}{R_l}\right)^{-H_l} \left(1 + \left(\frac{x' - m}{R_e}\right)\right)^{-H_e} dx\, dx'$$

$$\text{(eq. 5)}$$

The only difference between equation-3 and equation-5 is that the upper limit in the outer integral of the latter is c, not ∞. Of course, we do not have complete confidence in this probability, because we have no information on what may or may not have been fossilized in the region of nondeposition. Thus our confidence in the probability given by equation-5 is the confidence we have that the later appearing taxon actually originated somewhere between its first appearance in the fossil record and c My below that point. This confidence is given by:

$$C_{(B_l \text{ above } B_e; \text{ truncation } c \text{ myr below } B_l)} = 1 - \left(1 + \frac{c}{R_l}\right)^{-(H_l - 1)} \qquad \text{(eq. 6)}$$

Equation-6 is the same as equation-16 in Strauss and Sadler 1989. Accordingly, given a hiatus c My below the observed base of the earlier appearing taxon, equation-4 now becomes:

$$P(B_l \text{ below } B_e; \text{ truncation } c \text{ myr below } B_e) =$$

$$\left(\frac{H_e - 1}{R_e}\right)\left(\frac{H_l - 1}{R_l}\right) \int_m^c \int_{x'}^c \left(1 + \frac{x}{R_l}\right)^{-H_l} \left(1 + \left(\frac{x' - m}{R_e}\right)\right)^{-H_e} dx\, dx'$$

$$\text{(eq. 7)}$$

Our confidence in this probability is given by:

$$C_{(B_l \text{ below } B_e; \text{ truncation } c \text{ myr below } B_e)} = 1 - \left(1 + \frac{c - m}{R_e}\right)^{-(H_e - 1)} \qquad \text{(eq. 8)}$$

Establishing the Order of Extinction from Fossil Data

Equation-1 through equation-8 are easily adapted to cases of extinction. The probability that for the earlier disappearing taxon the stratigraphic position of the true but unknown top (time of extinction, T_e) is lower (older) than the true top of the later disappearing taxon (T_l), i.e., $P(T_e$ below $T_l)$, is given by:

$$P(T_e \text{ below } T_l) = 1 - \left(1 + \frac{m}{R_e}\right)^{-(H_e - 1)} +$$

$$\left(\frac{H_e - 1}{R_e}\right)\left(\frac{H_l - 1}{R_l}\right) \int_{Jm}^{\infty} \int_{Jm}^{x'} \left(1 + \frac{x}{R_e}\right)^{-H_e} \left(1 + \left(\frac{x' - m}{R_l}\right)\right)^{-H_l} dx \, dx'$$

(eq. 9)

where R_e, H_e, R_l, and H_l designate the observed stratigraphic ranges and numbers of fossil horizons of the earlier and later disappearing taxa respectively, and m is the gap between the times of disappearance of the two taxa. Equation-9 has the same properties as equation-3, the only difference is that terms R_e and H_e have been switched with R_l and H_l. Note that m is still the mismatch in the endpoints of the two taxa, but now stratigraphic distance is measured from the top of the earlier disappearing taxon, and increases up section.

The probability that the later disappearing taxon really became extinct earlier than the earlier disappearing taxon $P(T_e$ above $T_l)$ is given by:

$$P(T_e \text{ above } T_l) =$$

$$\left(\frac{H_e - 1}{R_e}\right)\left(\frac{H_l - 1}{R_l}\right) \int_{Jm}^{\infty} \int_{x'}^{\infty} \left(1 + \frac{x}{R_e}\right)^{-H_e} \left(1 + \left(\frac{x' - m}{R_l}\right)\right)^{-H_l} dx \, dx'$$

(eq. 10)

Determining Whether the Order of Appearance of Taxa is an Effect of Hiatuses in the Fossil Record

The Signor-Lipps effect, coined by Raup (1986), refers to the fact that with incomplete preservation (random range truncations) sudden extinction events will appear as gradual declines in the fossil record (Signor and Lipps 1982). Similarly, if a number of taxa exist prior to the opening of a taphonomic window conducive to their preservation, then the first appearances of those taxa in the fossil record may be smeared out—giving the appearance of a sequential origination, even though all were extant prior to the time of opening of the taphonomic window. Thus all *Metrarabdotos* species known from rocks younger than 8 Ma first appear in the fossil record within 1 My of the opening of the taphonomic window conducive to fossilization at 8 Ma. It is thus possible that all these taxa originated before the opening of the taphonomic window at 8 Ma, and that their order of appearance in the fossil record is just an effect of the hiatus in the rock record that ended at 8 Ma coupled with the incompleteness of the fossil record.

Springer (1990), using Strauss and Sadler's (1989) confidence intervals, has provided a way of distinguishing between sudden extinctions that appear gradual (owing to the Signor-Lipps effect) and true gradual extinctions. The same approach may be used to distinguish between "sudden originations" that simply appear sequential because of the incompleteness of the fossil record (the Sppil-Rongis effect) and true cases of sequential origination. The method is straightforward, and is applied in two steps.

First, for each taxon the confidence level (C) corresponding to the confidence interval (r/R) that reaches from the base of the known stratigraphic range to the time of opening of the taphonomic window is calculated, using the equation derived by Strauss and Sadler (1989):

$$C = 1 - \left(1 + \frac{r}{R}\right)^{-(H-1)} \qquad \text{(eq. 11)}$$

where r is the range extension necessary to bridge the gap between the base of the stratigraphic range and the time of opening of the taphonomic window; R is the stratigraphic range; and H is the number of fossil horizons. (Note that this equation is equivalent to equation-1 and equation-6.)

The second part of the method pertains to the null hypothesis that all taxa were extant prior to the opening of the taphonomic window. If the null hypothesis is correct, the confidence levels calculated in the first step are expected to be uniformly distributed (figure 7.3). This hypothesis can be tested by using the Kolmogorov-Smirnov Goodness of Fit Test (Hogg and Tanis 1977, pp. 281–286, 431), or any other suitable statistical test.

Figure 7.3   Confidence intervals used to distinguish "instantaneous originations" (left side) from sequential originations (right side), in light of the incompleteness of the fossil record. A: hypothetical taxa shown with true stratigraphic ranges. B: observed stratigraphic ranges that result from the random selection of three fossil horizons from the true ranges shown in (A). The number at the base of each range is the confidence level that corresponds to the con-

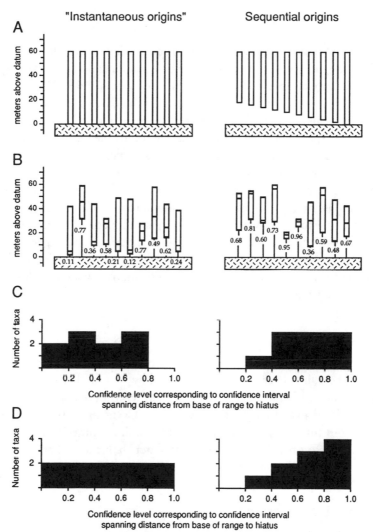

fidence interval that spans the distance from the base of the fossil range to the hiatus at 0 meters (equation-11). C: frequency distributions of the confidence levels shown in (B). D: the expected frequency distribution of confidence levels for an "instantaneous origination"—which is a uniform distribution (left column). If the taxa did indeed originate sequentially, then the data will "pile up" on the right side of the histogram (right column). The Kolmogorov-Smirnov Goodness of Fit Test, performed on cumulative frequency distributions, may be used to test whether the observed distributions deviate significantly from a uniform distribution. For the distributions shown in (C), the left side cannot be distinguished from a uniform distribution ($p > 0.2$), but the distribution on the right deviates significantly from a uniform distribution ($0.05 < p < 0.1$).

The importance of this two-step approach is that it formally recognizes the role of sample size in determining the significance of stratigraphic patterns. For example, consider species A for which the 95% confidence interval spanned the entire distance between the base of its range and the top of a hiatus. We would conclude that there is only a one-in-twenty chance that species A originated *below* the top of the hiatus. However, if twenty species had the bases of their ranges truncated by the hiatus (i.e. all species originated below the top of the hiatus), we would expect approximately one of those species to have a gap between the top of the hiatus and its first appearance to correspond to a confidence level greater than or equal to 0.95. Maybe species A just happens to be that species, and it really did originate below the top of the hiatus. Springer's (1990) approach can be used to evaluate the spread of confidence levels for an ensemble of taxa, but it is relatively uninformative when each taxon is examined in isolation.

Levels of Significance for Statistical Tests

It is conventional to reject null hypotheses only if $p < 0.05$. For the methods used here, however, a probability that low is viewed as too stringent, and p values of between 0.1 and 0.2 were considered significant. For example, if the probability $P(B_1$ above $B_e) = 0.8$ (i.e., $p = 0.2$), then even though conventionally one cannot reject the null hypothesis that the later appearing taxon really did not originate later than the earlier appearing taxon, there is still a one-in-five chance that it did.

Random Fossilization?

Both methods just described—(1) the test for a significant difference in the times of first appearances between pairs of taxa in the fossil record and (2) the test for distinguishing sequential from sudden originations of ensembles of taxa—require an assumption that the fossil horizons for each taxon are distributed uniformly within the observed stratigraphic range. If, however, the distribution of fossil horizons deviates significantly from a uniform distribution, then (following Strauss and Sadler 1989), these methods may only be used to gain a rough, or base line, estimate of the significance of order of appearance of taxa in the fossil record. I used the Kolmogorov-Smirnov Goodness of Fit Test (Hogg and Tanis

1977, p. 281–286, 431) to test whether the assumption of uniformity was indeed consistent with the bryozoan data I wished to analyze. The test was applied to cumulative frequency distributions, not directly on the type of frequency distribution shown in figures 7.3C and 7.3D.

In the remainder of this paper, the methods I have described are applied to the fossil record of the bryozoan species belonging to the genus *Metrarabdotos*. Cheetham (1986) proposed ancestor-descendant relationships between species pairs based on the order in which nearest neighbors in morphospace first appear in the fossil record. I evaluate Cheetham's stratophenetic hypotheses stratigraphically, using the equations described here. In addition, noting the hiatus in the rock record immediately prior to a major appearance of species at 8 Ma, I evaluate the possibility that this apparent evolutionary radiation at 8 Ma is just an effect of the incompleteness of the fossil record coupled with the hiatus in the rock record.

Ancestor-Descendant Relationships Among Species of *Metrarabdotos*

Cheetham (1986) used a stratophenetic approach to assess the phylogenetic relationships among the species of *Metrarabdotos*. Species were placed in a morphospace, and each pair of nearest neighbors was assumed to be related in an ancestor-descendant relationship. For each pair the first appearing species was assumed to be ancestral, and the later appearing descendant. Equation-5 may be used to calculate the probability that later appearing species did indeed originate later than the earlier appearing species, as predicted by the hypothesized relationships. It should be noted that my own study is not concerned with the various and difficult issues that surround the identification of ancestors and descendants in the fossil record; the analysis is simply concerned with determining how strongly the stratigraphic data alone support the order of origin of *Metrarabdotos* species predicted by Cheetham's hypotheses of ancestor-descendant relationship.

The mathematical equations described in the first half of this paper were based on the assumption of randomly placed fossil horizons. Table 7.1 shows the results of the statistical test of the validity of the assumption of randomly placed fossil horizons for each species. For the well-preserved specimens, 9 of the 11 species have distributions of fossil horizons that

Table 7.1

Are Fossil Horizons of Metrarabdotos *Species Randomly Distributed?*

Metrarabdotos Species	Well-Preserved Fossils Only	All Fossil Horizons
n. sp. 10	—	—
n. sp. 9	—	**
auriculatum	—	—
colligatum	***	***
n. sp. 5	***	**
n. sp. 6	—	—
n. sp. 7	—	***
n. sp. 8	—	—
lacrymosum	—	—
n. sp. 3	—	—
n. sp. 4	—	—

NOTE: Null hypothesis of random distribution tested with Kolmogorov-Smirnov Goodness of Fit Test. Null hypothesis rejected at p values indicated: (***) = $p < 0.01$; (**) = $0.01 < p < 0.05$; (*) = $0.05 < p < 0.1$; (−) = $p > 0.1$.

cannot be distinguished from a uniform distribution at $p > 0.1$. When the less well preserved specimens are included, 7 of the 11 species have distributions that are indistinguishable from a uniform distribution at $p > 0.1$.

Table 7.2 shows the probability for each hypothesized ancestor-descendant pair that the putative descendant did indeed originate later than the putative ancestor. These values are given for the well-preserved specimens only, as well as for all fossil horizons of *Metrarabdotos*. Where either the putative ancestor or descendant (or both) have nonrandomly distributed fossil horizons, the probabilities are given in brackets, and the values should be regarded only as first order estimates of the true but unknown probabilities.

Also given are the confidence values for the calculated probabilities. Most of these values are less than one, reflecting the absence of rock prior to 8 Ma in the Dominican Republic sections. It is clear that the hiatus ending 8 Ma has some effect on the pattern of originations of the *Metrarabdotos* species.

Even at $p > 0.2$, only a few of the hypothesized ancestor-descendant relationships find significant stratigraphic support (table 7.2, figure 7.4). This is not to say that the stratigraphic data contradict the other hypothesized relationships, just that they provide no special support. The hy-

Table 7.2

Stratigraphic Support for the Hypothetical Ancestor-Descendant Relationships Between Species of Metrarabdotos

Putative Ancestor	Putative Descendant	Well-Preserved Fossils Only		All Fossil Horizons	
		P (Descendant Younger than Ancestor)	Confidence in Probability	P (Descendant Younger than Ancestor)	Confidence in Probability
n. sp. 9	n. sp. 10	**1.00**	1.00	[1.00]	[1.00]
auriculatum	n. sp. 9	**0.90**	0.90	[0.98]	[0.98]
colligatum	auriculatum	[0.00]	[0.00]	[0.00]	[0.00]
colligatum	n. sp. 5	[0.78]	[0.78]	[0.75]	[0.75]
n. sp. 5	n. sp. 6	[0.92]	[0.94]	[0.75]	[0.93]
n. sp. 6	n. sp. 7	0.05	0.77	[0.76]	[0.89]
n. sp. 7	n. sp. 8	**0.83**	0.89	[0.95]	[1.00]
lacrymosum	n. sp. 3	0.36	0.52	0.69	0.79
n. sp. 3	n. sp. 4	0.60	0.77	**0.93**	0.98

NOTE: Probabilities given by equation-5, confidence in these probabilities by equation-6. Square brackets indicate cases where either the ancestor or the descendant, or both, have distributions of fossil horizons that differ from random (see table 7.1). In these cases the probabilities should be viewed as first-order estimates only. Values in bold indicate hypotheses finding the strongest support.

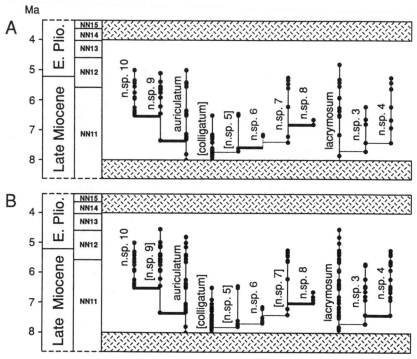

Figure 7.4 Fossil record and stratigraphic support for hypothetical ancestor-descendant relationships among species of *Metrarabdotos*. Stippled regions indicate times of little or no rock. Note that some fossil horizons are sufficiently close that they are superimposed (the stratigraphic data may be found in appendix A). No significance should be attached to the horizontal distance between species, though adjacent species were adjacent in morphospace (Cheetham 1986). Names in square brackets [] indicate species with nonrandom distribution of fossil horizons (table 7.1). Hypothetical ancestor-descendant relationships (Cheetham 1986) are shown by the horizontal lines, and those finding significant stratigraphic support are indicated by the heavy horizontal lines, though those involving taxa with nonrandom distributions of fossil horizons should be viewed as baseline estimates only. A: well-preserved specimens only. B: all horizons yielding *Metrarabdotos* fossils.

potheses finding significant stratigraphic support are broadly congruent between the analyses of the well-preserved and complete data sets, even though for almost all ancestor-descendant species pairs in the complete data set at least one of the species has a nonrandom distribution of fossil horizons (table 7.1, figure 7.4).

No Radiation of Metrarabdotos Species at 8 Million Years Ago?

Many of the eleven *Metrarabdotos* species that appear in the fossil record only after 8 Ma first appear relatively close to the opening of the taphonomic window at 8 Ma (figure 7.4). The results of the tests for the possibility that all, or most, of these species of *Metrarabdotos* actually originated prior to 8 Ma are shown in tables 7.3 and 7.4 for the well-preserved-specimen and all-specimen data sets respectively. In both the cases the hypothesis that all eleven species originated prior to 8 Ma is rejected (with the caveat that a number of species have nonrandom distributions of fossil horizons).

Even though the analyses indicate that all eleven species were not yet extant by 8 Ma, inspection of the stratigraphic data (figure 7.4) shows that it is unlikely that only the two putative ancestral species *M. colligatum* and *M. lacrymosum* had their origins prior to 8 Ma. To test this intuition, Springer's (1990) approach was applied to increasingly smaller subsets of the eleven species. There are no established criteria for choosing the order in which species should be eliminated from such an analysis. However, it seemed reasonable that species least likely to have originated prior to 8 Ma should be eliminated first. This probability (confidence, actually) was calculated using equation-6, and the results are shown in table 7.2 (for the species in the "Putative Descendant" column, the corresponding probability is given in the "Confidence in Probability" column). Tables 7.3 and 7.4 show the results of the analysis of which species most likely had their origins prior to 8 Ma.

Table 7.3

Tests of Null Hypotheses that All, or at Least Some, Species of Metrarabdotos *Were Extant Prior to the Opening of the Taphonomic Window 8 Ma, Based on Well-Preserved* Metrarabdotos *Fossils Only*

Metrarabdotos Species Analyzed	Reject Null Hypothesis
All	**
All Except:	
n. sp. 10	*
n. sp. 10, and n. sp. 6	—

NOTE: Each null hypothesis was tested using Kolmogorov-Smirnov Goodness of Fit Test. Null hypotheses rejected at p values indicated: $(**) = 0.01 < p < 0.05$; $(*) = 0.1 < p < 0.2$; $(-) = p > 0.2$.

Table 7.4

Tests of Null Hypotheses that All, or at Least Some, Species of Metrarabdotos *Were Extant Prior to the Opening of the Taphonomic Window 8 Ma, Based on Analysis of All* Metrarabdotos *Fossils*

Metrarabdotos Species Analyzed	Reject Null Hypothesis	
	All Taxa Analyzed	Nonrandom Species Eliminated
All	****	***
All Except:		
n. sp. 10	***	**
n. sp. 10⎱ n. sp. 8 ⎰	***	*
n. sp. 10⎱ n. sp. 8 ⎬ n. sp. 4 ⎰	**	—
n. sp. 10⎱ n. sp. 8 ⎬ n. sp. 4 ⎰ n. sp. 9	*	not done
n. sp. 10⎱ n. sp. 8 ⎬ n. sp. 4 ⎬ n. sp. 9 n. sp. 6 ⎰	—	not done

NOTE: Each null hypothesis was tested using Kolmogorov-Smirnov Goodness of Fit Test. In right column the four species with nonrandom distributions of fossil horizons (n. sp. 9, *M. colligatum*, n. sp. 5, and n. sp. 7, table 7.1) were eliminated from the analysis. Null hypothesis rejected at p values indicated: (****) $= p < 0.01$; (***) $= 0.01 < p < 0.05$; (**) $= 0.05 < p < 0.1$; (*) $= 0.1 < p < 0.2$; (−) $= p > 0.2$

Based on the analysis of the well-preserved fossils (table 7.3), it appears likely that n. sp. 10 and n. sp. 6 originated later than 8 Ma. Two species (*M. colligatum* and n. sp. 5) have nonrandom distributions of fossil horizons, but if these are eliminated from the analysis, one still cannot reject the null hypothesis that all but n. sp. 10 and n. sp. 6 had their origins prior to the opening of the taphonomic window at 8 Ma.

With the addition of the poorly preserved fossils, the analysis indicates that considerably more taxa (n. sp. 10, n. sp. 8, n. sp. 4, n. sp. 9, and n. sp. 6) had their origins after the opening of the taphonomic window 8 Ma. The discrepancy between the two analyses stems from the fact that

while there are considerably more fossil horizons when the poorly preserved fossils are added, very little range extension results from the additional data (see appendix A). The results of the analysis of the complete
data set may be compromised by the inclusion of the four species with
nonrandom distributions of fossil horizons. However, if these four taxa
(n. sp. 9, *M. colligatum,* n. sp. 5, and n. sp. 7) are omitted, one still finds
considerable evidence that several of the remaining species originated
more recently than 8 Ma (n. sp. 10, n. sp. 8, n. sp. 4). Four species remain
as candidates for origination prior to 8 Ma (n. sp. 6, n. sp. 3, *M. auriculatum* and *M. lacrymosum*), though with such a small sample size it
becomes difficult to reject any null hypothesis.

Overall, the analysis shows that only a few species probably originated
before the opening of the taphonomic window at 8 Ma. But given that
this type of analysis is designed to test the properties of ensembles of taxa,
the analysis is relatively uninformative about which species, in particular,
existed prior to 8 Ma.

Summary: Stratigraphy and the Evolution of *Metrarabdotos*

Only four of nine hypothesized ancestor-descendant relationships find
strong probabilistic-stratigraphic support (n. sp. 10 descending from n.
sp. 9; n. sp. 9 from *M. auriculatum*; n. sp. 8 from n. sp. 7; and n. sp. 4
from n. sp. 3). For the well-preserved fossils alone, one hypothesis of
ancestor-descendant relationship (n. sp. 7 from n. sp. 6) may be rejected
on stratigraphic grounds because the putative ancestor (n. sp. 6) first appears later than its putative descendant (n. sp. 7) (figure 7.4A); this hypothesis is not, however, rejected when the poorly preserved fossils are
added. It is important not to over-interpret these results; except for the
hypothesized relationship between n. sp. 6 and n. sp. 7 in the well-
preserved data set (figure 7.4A), none of the hypothesized relationships
may be rejected on stratigraphic grounds; the analysis simply shows that
five of nine hypotheses find no *special* stratigraphic support.

The hypothesis that all *Metrarabdotos* species were extant prior to 8
Ma is soundly rejected. However, there is evidence that perhaps several
species may have originated prior to that time. The results are tentative,
owing to the absence of criteria for sequentially eliminating taxa from the
analysis, because of the difficulty associated with species having nonran-

dom distributions of fossil horizons and because of a relative lack of taxa and fossil horizons.

A striking feature of Cheetham's (1986) "stratophenogeny"—that is, his phylogenetic tree—is that at every speciation event there is no morphologic change in the ancestral species. However, this finding may be an artifact of a literal reading of the fossil record. Consider a bifurcating speciation event where both newly originated species differ from the ancestor but show stasis after their originations. With incomplete fossilization one of these new taxa is likely to appear before the other in the fossil record. If this taxon were mistaken as the ancestor of the other species, then this "ancestor" would seem to have remained in stasis during the origin of the later appearing form. Certainly my own analysis indicates that (if nearest neighbors in morphospace are closest relatives) some descendant species originated while the ancestor remained in stasis. But for species not showing significant stratigraphic support for the hypothesized ancestor-descendant relationship, there remains the possibility that speciation involved morphological divergence in more than just one species at the time of speciation.

A conclusion that perhaps half the 11 species of *Metrarabdotos* first appearing in the rock record at 8 Ma actually originated earlier than 8 Ma is not unreasonable from a stratigraphic perspective. There is, after all, essentially no rock record for more than half the Neogene in the Dominican Republic (Saunders, Jung, and Biju-Duval 1986), and it is perhaps unlikely that the majority of speciation events known to have occurred within the genus *Metrarabdotos* would fortuitously have occurred at times of a good rock record. Further, two species that have the same time of origin (due to a bifurcating cladogenic event) or the same apparent time of origin (because both species existed prior to the opening of an appropriate taphonomic window) are likely, with incomplete fossilization, to appear sequentially in the fossil record. If a sequence of first appearances is suspected of being a consequence of incomplete preservation, then the stratigraphic data for any pair of species should have two properties: (1) The species with the poorer fossil record (measured in number of fossil horizons per unit time) should (statistically) appear later than the taxon with the richer fossil record. (2) The mismatch in their times of first appearance should have approximately the same size as the difference in the median gap size of the two taxa.

Visual inspection of the *Metrarabdotos* fossil record shows that these two conditions apply for a number of species pairs (figure 7.4), though by no means all (e.g., n. sp. 10 and n. sp. 9). It is not unreasonable then to suppose that several species had their origins during one of the major hiatuses in the rock record of the Dominican Republic. However, while these rules of thumb have their value, it is only with application of methods such as those developed here that we can gain confidence in the significance or reality of the order of appearance (or disappearance) of taxa in the fossil record.

Dan Fisher (1980, 1988, 1991, 1992) has argued that stratigraphic data should play a role in phylogenetic analysis, and has suggested that hypotheses of relationship should incur a parsimony debt not only with respect to conflicting distributions of morphologic attributes but also with respect to conflicting stratigraphic data. I too have argued that stratigraphic data may be used to help evaluate hypotheses of relationship (Marshall 1990a,b). The methods developed in this paper may be applied to the practice of stratocladistics, providing a way of evaluating the significance of conflicting stratigraphic distributions, though obviously only for groups for which detailed stratigraphic data may be obtained.

Appendix A

Stratigraphic Positions of *Metrarabdotos* Fossil Horizons

n. sp. 10					n. sp. 8
150 [3.85]	49 6.40		auric.	181 6.35	[3.85]
8 5.00	189 6.45	5.55	5 [1.85]	6.40	94 6.65
5.30	56 6.50	5.55	13 [3.3]	53 6.45	103 6.80
76 5.35		32 5.60	14 [3.35]	51 6.45	6.90
81 5.40		35 5.65	4.80	62 6.50	6.90
25 5.40	n. sp. 9	5.65	4.80	194 6.50	7.00
28 5.55	152 [3.85]	39 5.75	19 5.00	7.05	
5.65	4.55	41 5.90	5.15	108 7.15	
40 5.75	5.00	5.95	23 5.40	7.15	
42 5.90	21 5.10	5.95	26 5.45	120 7.20	
175 5.95	5.15	204 6.20	27 5.55	65 7.25	
176 6.25	24 5.40	50 6.45	5.55	69 7.30	
6.25	5.45	70 7.35	5.55	7.35	
45 6.35	29 5.55		5.65	140 7.80	
			6.25	145 8.00	

collig.

SPL	Age
197	6.50
	6.90
	7.05
111	7.15
	7.15
117	7.20
66	7.25
122	7.30
134	7.40
	7.45
	7.50
	7.50
	7.55
214	7.60
125	7.60
	7.60
	7.60
208	7.65
218	7.70
	7.70
	7.70
136	7.75
221	7.75
	7.75
139	7.80
	7.80
128	7.85
	7.85
	7.85

SPL	Age
130	7.90
225	7.90
143	7.90
7	8.00
144	8.00

n. sp. 5

SPL	Age
55	6.45
52	6.45
63	6.50
60	6.50
57	6.50
123	7.60
127	7.70
126	7.70
	7.75
	7.75
	7.80

n. sp. 6

SPL	Age
112	7.15
115	7.20
	7.50
	7.60
	7.65
	7.70
	7.70

n. sp. 7

SPL	Age
72	5.25
	5.30
77	5.35
	5.40
83	5.45
84	5.60
	5.65
89	6.20
119	7.20
210	7.40

lacrym.

SPL	Age
6	[2.5]
	[3.4]
153	[3.85]
	4.55
17	4.80
	5.00
	5.25
74	5.30
	5.35
80	5.40
30	5.55
	5.60
	5.65

SPL	Age
44	5.90
	5.95
179	6.25
46	6.35
	6.40
	6.45
	6.45
	6.45
64	6.50
58	6.50
61	6.50
	6.50
97	6.55
	6.55
	6.80
	6.90
	6.90
	7.00
113	7.15
	7.15
	7.20
	7.30
	7.40
	7.45
	7.50
	7.80
129	7.85
	7.85
	7.90

n. sp. 3

SPL	Age
154	[3.85]
	5.70
	5.70
	5.75
92	6.20
	6.45
	6.50
102	6.65
106	6.80
	6.90
132	7.40
209	7.70
	7.70

n. sp. 4

SPL	Age
73	5.25
	5.35
82	5.40
87	5.60
	5.65
	5.70
	5.75
	5.90
90	6.20
	6.45
	6.50
	6.55
100	6.65
	6.80
205	6.90
	6.90
217	7.40
	7.40

Note: For each *Metrarabdotos* species, left column gives SPL numbers, which correspond to the well-preserved specimens used by Cheetham (1986, 1987), and by Cheetham and Hayek (1988). Horizons without SPL numbers correspond to *Metrarabdotos* fossils too poorly preserved to be included in those morphological studies. Right column gives ages, in Ma, of each *Metrarabdotos* fossil. These ages reflect revisions made since those publications. Horizons with the same age are known, for the most part, from different stratigraphic horizons from (usually) the same section. The precision on the age assignments is no better than 0.05 My. All data from Cheetham (pers. comm. 1992). Entries in brackets are younger than 4 Ma and were excluded from the analyses. Abbreviations: auric = *M. auriculatum*; collig = *M. colligatum*; lacrym = *M. lacrymosum*.

Appendix B

Derivation of Equations

Strauss and Sadler (1989) derived an equation (their equation-16) for calculating confidence intervals on one endpoint of stratigraphic ranges:

$$P(0 < B \le m) = 1 - \left(1 + \frac{m}{R}\right)^{-(H-1)} \qquad \text{(eq. A1)}$$

where R is the observed stratigraphic range, H the number of fossil horizons, and m some distance below the base of the range. This equation may be viewed as a distribution function, $F(x)$, for the probability that the interval of width m below the base of the range includes the true base (B), or time of origin, of a taxon. The corresponding density distribution function, $f(x)$, may be found by noting that the following holds (Kreyszig 1979, p. 866):

$$P(0 < B \le m) = F(m) - F(0) = \int_0^m f(v)\, dv \qquad \text{(eq. A2)}$$

where v is a dummy variable. Hence:

$$\frac{d[F(x)]}{dx} = f(x) \qquad \text{(eq. A3)}$$

Thus by differentiating equation-A1, the density distribution function, $f(x)$, may be found:

$$f(x) = \left(\frac{H-1}{R}\right)\left(1 + \frac{x}{R}\right)^{-H} \qquad \text{(eq. A4)}$$

and equation-A1 (Strauss and Sadler's equation-16) may be rewritten in the following generalized form:

$$P(a < B \le m) = \left(\frac{H-1}{R}\right)\int_a^m \left(1 + \frac{x}{R}\right)^{-H} dx \qquad \text{(eq. A5)}$$

where $P(a < B \le m)$ is the probability that the interval between a My and m My below the base of the known stratigraphic range includes the true base of the range.

For two independent random variables (one variable for the true time of origin of each taxon), the two-dimensional distribution function is given by (Kreyszig 1979, pp. 885–889):

$$P(a_1 < B_l \leq b_1, a_2 < B_e \leq b_2) = \int_{a_2}^{b_2} \int_{a_1}^{b_1} f_l(x) f_e(x') dx\, dx' \quad \text{(eq. A6)}$$

where B_l is the true time of origin of the later appearing taxon, B_e is the true time of origin of the earlier appearing taxon. $f_l(x)$ is the probability density function for the position of the true time of origin of the later appearing taxon, and from equation-A4 is given by:

$$f_l(x) = \left(\frac{H_l - 1}{R_l}\right)\left(1 + \frac{x}{R_l}\right)^{-H_l} \quad \text{(eq. A7)}$$

where H_l and R_l are the number of fossil horizons and the stratigraphic range of the later appearing taxon. The probability density function for the true time of origin of the earlier appearing taxon, $f_e(x')$, is given by:

$$f_e(x') = \left(\frac{H_e - 1}{R_e}\right)\left(1 + \left(\frac{x' - m}{R_e}\right)\right)^{-H_e} \quad \text{(eq. A8)}$$

where H_e and R_e are the number of fossil horizons and stratigraphic range of the later appearing taxon. Note that if $H_l = H_e$, and if $R_l = R_e$, then:

$$f_l(x) = f_e(x + m) \quad \text{(eq. A9)}$$

reflecting the mismatch of m My in the bases of the two taxa's stratigraphic ranges. (Springer and Lilje [1988] proceeded in much the same way to derive an equation for calculating the same probability, but mistakenly set $f_l(x) = f_e(x)$. The other major difference between their approach and the one used here is that they used the exponential model for the density distribution for gap sizes [see Strauss and Sadler 1989 for a discussion of this approach]).

To calculate the probability that the later appearing taxon actually arose later than the earlier appearing taxon in the region where both could have originated [$P(B_l$ above B_e; $B_l > m)$], we need to select appropriate values of $a_1, a_2, b_1,$ and b_2 for equation-A6. For some arbitrary point x', we need the probability that the interval below m ($a_1 = m$), but above x' ($b_1 = x'$) includes the true base of the later appearing taxon, B_l (figure 7.1B). For the earlier appearing taxon we need the probability that the interval below x' includes its true base. Thus for any specific value of x', $a_2 = x'$, and $b_2 = \infty$. But we require the probability for all possible values of x'. This is accomplished by allowing x' to vary from m to in the outer integral; therefore $a_2 = m$ and

$b_2 = \infty$. Thus, using these limits and substituting equation-A7 and equation-A8 into equation-A6, we obtain equation-2 given in the text:

$P(B_l \ above \ B_e; \ if \ B_l > m) =$

$$\left(\frac{H_e - 1}{R_e}\right)\left(\frac{H_l - 1}{R_l}\right)\int_m^\infty \int_m^{x'} \left(1 + \frac{x}{R_l}\right)^{-H_l} \left(1 + \left(\frac{x' - m}{R_e}\right)\right)^{-H_e} dx \ dx'$$

(eq. A10)

Note that this equation is quite general. If one wished to calculate the probability that the first appearing taxon really originated later than the later appearing taxon, then $a_1 = x'$, and $b_1 = \infty$, and $a_2 = m$, and $b_2 = \infty$, (see equation-4 in the text). Similarly, if there is a truncation event c My below the base of the later appearing taxon, then $b_2 = c$, rather than ∞ (see equation-5 in the text).

Appendix C

Mathematica and Double Integrals

To solve double integrals with the computer package *Mathematica* the following general format is used (Wolfram 1991, p. 106):

$$\text{NIntegrate}[f, \{x, x_{min}, x_{max}\}, \{y, y_{min}, y_{max}\}] \qquad \text{(eq. A11)}$$

which is equivalent to:

$$\int_{x_{min}}^{x_{max}} dx \int_{y_{min}}^{y_{max}} dy \ f \qquad \text{(eq. A12)}$$

where f is the integrand. The integrand is integrated with respect to y first, then to x. Thus, our x is *Mathematica*'s y and our x' is *Mathematica*'s x. The *Mathematica* code for equation-2 and equation-A10 is:

$$\text{NIntegrate}[(((H_e - 1) \ (H_l - 1) \ (1 + (y/R_l)) \ \wedge \ -H_l \ (1 +$$
$$(x - m)/R_e) \ \wedge \ -H_e) \ /(R_e \ R_l)), \{x, m, \text{Infinity}\}, \{y, m, x\}] \qquad \text{(eq. A13)}$$

For any particular problem, the specific values for the variables are typed in and the solution will be printed to the screen shortly after pushing the enter key. If the variables are assigned the following values, $H_e = 3$, $H_l = 10$, $R_e = 0.4$, $R_l = 1.1$, $m = 0.2$, then equation-A13 becomes:

NIntegrate[(((2) (9) (1 + (y/1.1)) ^ $-$ 10 (1 +
   (x $-$ 0.2)/0.4) ^ $-$ 3)/(0.4 1.1)),{x, 0.2, Infinity}, {y, 0.2, x}]   (eq. A14)

and the value of this double integral computed by *Mathematica* is 0.135812.

Acknowledgments   I thank Alan Cheetham for generously providing unpublished stratigraphic data. This paper has benefited greatly from discussions with Gerhard Oertel, Bruce Runnegar, and Stefan Bengtson. The final manuscript was improved with suggestions from Doug Erwin, Gerhard Oertel, Peter Sadler, and Connie Barlow. This work has been partially supported by NSF grant no. EAR-9258045, support that is greatly appreciated. Acknowledgment is also made to the donors of the Petroleum Research Fund, administered by the American Chemical Society, for partial support of this research.

References

Cheetham, A. H. 1986. Tempo of evolution in a Neogene bryozoan: Rates of morphologic change within and across species boundaries. *Paleobiology* 12:190–202.

———. 1987. Tempo of evolution in a Neogene bryozoan: Are trends in single morphologic characters misleading? *Paleobiology* 13:286–296.

Cheetham, A. H. and L. C. Hayek. 1988. Phylogeny reconstruction in the Neogene bryozoan *Metrarabdotos*: A paleontologic evaluation of methodology. *Historical Biology* 1:65–83.

Fisher, D. C. 1980. The role of stratigraphic data in phylogenetic inference. *Geological Society of America Abstracts with Programs* 12:426.

———. 1988. Stratocladistics: Integrating stratigraphic and morphologic data in phylogenetic inference. *Geological Society of America Abstracts with Programs* 20:A186.

———. 1991. Phylogenetic analysis and its application in evolutionary paleobiology. In N. Gilinsky and P. Signor, eds., *Analytical Paleobiology. Short Courses in Paleontology* 4:103–122.

———. 1992. Stratigraphic parsimony. In W. P. Maddison and D. R. Maddison. *MacClade: Analysis of Phylogeny and Character Evolution*, Version 3, pp. 124–129. Sunderland, Mass.: Sinauer.

Hogg, T. V. and E. A. Tanis. 1977. *Probability and Statistical Inference*. New York: Macmillan.

Kreyszig, E. 1979. *Advanced Engineering Mathematics*. New York: John Wiley and Sons.

Marshall, C. R. 1990a. Confidence intervals on stratigraphic ranges. *Paleobiology* 16:1–10.

———. 1990b. The fossil record and estimating divergence times between lineages: Maximum divergence times and the importance of reliable phylogenies. *Journal of Molecular Evolution* 30:400–408.

Raup, D. M. 1986. Biological extinction in earth history. *Science* 231:1528–1533.

Saunders, J. B., P. Jung, and B. Biju-Duval. 1986. Neogene Paleontology in the Northern Dominican Republic. 1. Field Surveys, Lithology, Environment, and Age. *Bulletins of American Paleontology* 89(323):1–79.

Signor, P. W., and J. H. Lipps. 1982. Sampling bias, gradual extinction patterns, and catastrophes in the fossil record. In L. T. Silver and P. H. Schultz, eds., *Geological Implications of Large Asteroids and Comets on the Earth. Geological Society of America, Special Paper* 190:291–296.

Springer, M. S. 1990. The effect of random range truncations on patterns of evolution in the fossil record. *Paleobiology* 16:512–520.

Springer, M. and A. Lilje. 1988. Biostratigraphy and gap analysis: The expected sequence of biostratigraphic events. *Journal of Geology* 96:228–236.

Strauss, D. and P. M. Sadler. 1989. Classical confidence intervals and Bayesian probability estimates for ends of local taxon ranges. *Mathematical Geology* 21:411–427.

Wolfram, S. 1991. *Mathematica*. Redwood City: Addison-Wesley.

III

Macroevolutionary Issues

8

Phylogeny, Diversity History, and Speciation in Paleozoic Bryozoans

Robert L. Anstey and Joseph F. Pachut

This paper is an initial attempt at presenting a macroevolutionary context for speciation in Paleozoic bryozoans. It includes a family-level phylogenetic analysis, delineates generic diversity patterns within suborders, and integrates macroevolutionary patterns with known speciation history.

The Nexus of Phylogeny, Diversity, and Speciation

Early Ordovician speciation is inferred to have produced significantly greater quantum levels of morphological evolution than at any subsequent time in the Paleozoic. All the higher taxa of Paleozoic bryozoans appeared during the Early Ordovician, and were not built up gradually through an accumulation of small differences in lower-level taxa. Most of the family branching and family-level apomorphies appeared rapidly in the initial radiation of the phylum, and without the benefit of extensive speciation. The Ordovician radiation of bryozoans, therefore, is strikingly parallel to the Cambrian metazoan radiation in this respect (Valentine and Erwin 1987; Foote and Gould 1992). Post-radiation speciation produced progressively decreasing morphological and taxonomic distances. Younger epeiric-sea species that were both long-lived and abundant display dynamic (oscillatory) stasis.

Our study has been facilitated by treatise-level revisions of genera and families within most higher taxa during just the last decade, by recent comprehensive compilations of species in the Ordovician and Silurian by Tuckey (1988, 1990a, 1990b) and in the Devonian by Horowitz and Pachut (1993), and by the availability of improved software for phylogenetic analysis. This study is targeted on all Paleozoic stenolaemate bryozoans, a clade distinguished by tubular zooecia. We excluded 14 genera of nonstenolaemates, namely, 6 genera of Paleozoic gymnolaemates (boring ctenostomes) and 8 genera of hederellines, a colonial group of uncertain zoological affinity. The target taxon minimally includes an estimated 5,200 species, 476 genera, 60 families, and 11 suborders. For comparison, Recent bryozoans include 5,330 species, of which only 700 are stenolaemates (Horowitz and Pachut 1993). Bryozoans represent a significant portion of the Paleozoic marine fossil record, and were co-dominant with brachiopods in epeiric sea benthic communities. Extant bryozoans are only rarely a dominant component of modern benthic marine communities.

We have attempted phylogenetic analysis first of the 11 stenolaemate suborders, using only 26 characters. Next we analyze the 60 Paleozoic stenolaemate families, using 58 two-state and additively coded multistate characters. We have found that the family-level cladograms can be restructured easily by the choices of generic exemplars within each family. Families are not defined by constant states in all of the 58 characters, even in those conventionally regarded as diagnostic. Many diagnostic-type characters also appear in the same or similar states in lineages conventionally considered to be widely divergent. Because simultaneous analysis of all 476 genera is not possible at this time (for software limitations as well as the immensity of the task), we have relied upon a combination of generic exemplars within each family. One analysis, which we believe reduces homoplasy due to multiple evolution of the same character states, relies upon the oldest known genus within each family. It, however, does not resolve the apparent paraphyly within both the bifoliate cryptostomes and the bifoliate cystoporates, which seems to be a problem largely connected with the multiple evolution of the bifoliate colony form. A second analysis utilizes younger and more typical or representative genera for each family, in an attempt to capture critical linkages to younger taxa.

With only a very small number of exceptions, we believe that the monophyly of most of the major higher taxa of Paleozoic bryozoans has been confirmed by our analyses. We make no claims for the finality of our phylogenetic analyses. That at least awaits a more detailed level of study, which should encompass all 476 genera. For the time being, however, these analyses provide working hypotheses of the cladogenesis and apomorphogenesis (evolution of derived character states) of Paleozoic bryozoans.

The diversity history of Paleozoic bryozoans provides a second essential context for analysis of the patterns, effects, and dynamics of speciation. The chronostratigraphic distributions of genera have been compiled from the published ranges of the 476 genera, and provide numbers of generic origins, generic extinctions, and standing diversities through 30 post-Cambrian Paleozoic stages or series. The use of absolute age dates permits estimations of rates of taxonomic evolution and extinction within clades. These data outline the major events in bryozoan history, such as adaptive radiations and mass extinctions, and distinguish episodes of high diversity flux from intervals of more slowly changing or equilibrium diversity.

Our assessment shows that the major higher taxon or novelty events took place during intervals of minimum generic diversity. Major generic increases accompanied or followed novelty radiations, and major decreases preceded higher-taxon extinctions. Therefore major macroevolutionary changes were strongly linked with levels of generic diversity and, by proxy, with species diversity.

Our main conclusion is that speciation has not remained a static phenomenon over the entire history of this major group of clades. Instead, it begins with the production of large effects (quantum evolution), and becomes increasingly damped over time.

Phylogeny of Bryozoan Families

Our initial phylogenetic analysis used each of the 11 Paleozoic stenolaemate suborders as operational taxonomic units, and coded each for its most frequent state in each of 26 two-state and multistate subordinal level characters (appendix A). This initial analysis is included here only as an introductory statement, as it was carried out several years ago using the branch and bound algorithm of PAUP 2.4; only two trees were found of 22 steps each, and a consistency index of 0.556. The tree illustrated (figure 8.1) demonstrates monophyly of the three major free-walled (double-walled) orders, linkage of trepostomes and cryptostomes as sister groups, and exclusion of the esthonioporines from the trepostomes. With the ex-

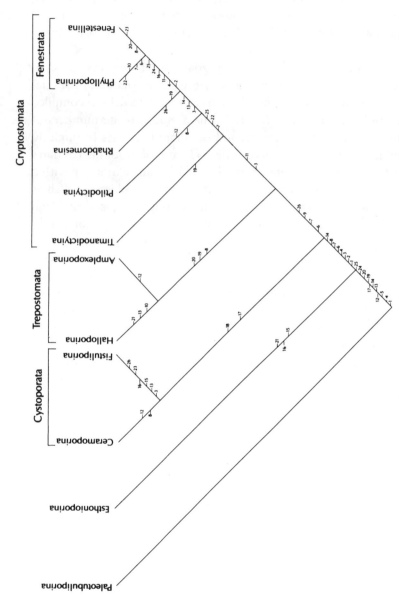

Figure 8.1 Cladogram of the 11 suborders of Paleozoic stenolaemate bryozoans, using characters listed in appendix A. Character numbers plotted to the right side of a branch indicate character state gains, whereas those plotted to the left side represent character state losses. The multiple appearance of numbered characters on either side of the branch indicates successive gains or losses within a multistate character.

ception of the position of the ceramoporines, the branching patterns in figure 8.1 have been entirely reproduced by the family-level analysis in figure 8.2.

Figure 8.2 is built from character codings for each family based on the oldest known genus within each family. Figures 8.2 and 8.3 were produced using PAUP 3.0, with maximum parsimony and random addition of taxa. Four of the 58 characters that initially displayed the highest levels of homoplasy were eliminated in subsequent analyses (appendix B).

Given the branching patterns in figure 8.2, the following relationships can be inferred:

- The Cryptostomata, Trepostomata minus Esthonioporina, Fistuliporina in part (6 families), and the Esthonioporina are ordinal-level clades.
- Subordinal clades include the Fenestrata, Amplexoporina, and Halloporina, with the Fenestellina and Phylloporinina as infraordinal clades.
- Subordinal paraclades within the Cryptostomata include the Rhabdomesina, Ptilodictyina, and Timanodictyina.
- The Cystoporata are polyphyletic, and the Fistuliporina are paraphyletic in part (5 families)

The Cryptostomata and Trepostomata are sister groups, and together are a sister group to the Fistuliporina. All three together are a sister group to the Esthonioporina (here excluded from Trepostomata). All four free-walled (double-walled) clades form a derived clade with respect to the Palaeotubuliporina, on which the cladogram is rooted. Two families, the Ulrichotrypellidae and Intraporidae, occupy cladogram positions inconsistent with their conventional systematic assignments (table 8.1). Likewise, the arrangement of the five basal families of fistuliporines is inconsistent with their conventional systematics. Intrinsic problems exist with using the oldest exemplar in each family. As primitive taxa, they often lack several of the derived character states that evolved later and that rank as diagnostic features of each family. Descendant families in many instances probably evolved from younger and more derived genera within ancestral families. Such critical linkages, especially for families first appearing in Silurian and younger strata, cannot be captured if Ordovician families are represented only by their oldest exemplars. Better resolution

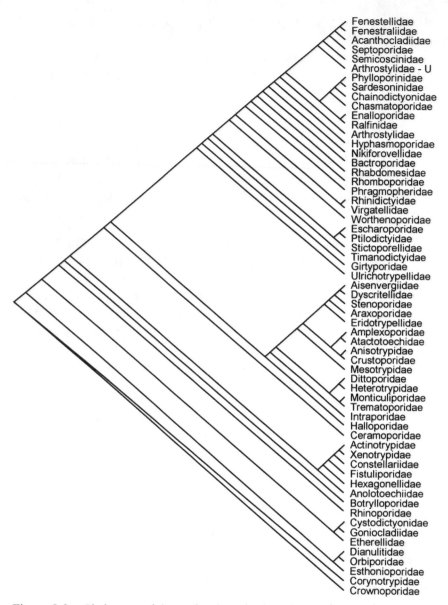

Fenestellidae
Fenestraliidae
Acanthocladiidae
Septoporidae
Semicoscinidae
Arthrostylidae - U
Phylloporinidae
Sardesoninidae
Chainodictyonidae
Chasmatoporidae
Enalloporidae
Ralfinidae
Arthrostylidae
Hyphasmoporidae
Nikiforovellidae
Bactroporidae
Rhabdomesidae
Rhomboporidae
Phragmopheridae
Rhinidictyidae
Virgatellidae
Worthenoporidae
Escharoporidae
Ptilodictyidae
Stictoporellidae
Timanodictyidae
Girtyporidae
Ulrichotrypellidae
Aisenvergiidae
Dyscritellidae
Stenoporidae
Araxoporidae
Eridotrypellidae
Amplexoporidae
Atactotoechidae
Anisotrypidae
Crustoporidae
Mesotrypidae
Dittoporidae
Heterotrypidae
Monticuliporidae
Trematoporidae
Intraporidae
Halloporidae
Ceramoporidae
Actinotrypidae
Xenotrypidae
Constellariidae
Fistuliporidae
Hexagonellidae
Anolotoechiidae
Botrylloporidae
Rhinoporidae
Cystodictyonidae
Goniocladiidae
Etherellidae
Dianulitidae
Orbiporidae
Esthonioporidae
Corynotrypidae
Crownoporidae

Figure 8.2   Cladogram of the 60 families of Paleozoic stenolaemate bryozoans, using characters listed in appendix B. Subordinal and ordinal assignments of all families are listed in table 8.1. Codings for each family were based upon the oldest known generic exemplar. Arthrostylidae-U = unilaminate arthrostylids; Arthrostylidae = radial arthrostylids.

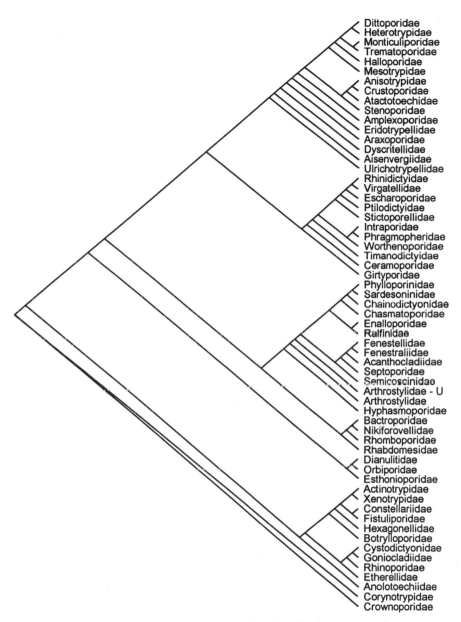

Figure 8.3 Alternate cladogram of the 60 families of Paleozoic stenolaemate bryozoans, using characters listed in appendix B. Codings for each family were based upon representative or typical generic exemplars. Note that rotation around the node joining the fenestrate-rhabdomesines with the trepostome-ptilodictyines will increase the resemblance in taxon order in the upper part of the tree to that in figure 8.2. Arthrostylidae-U = unilaminate arthrostylids; Arthrostylidae = radial arthrostylids.

Table 8.1

Conventional Systematic Assignments of Paleozoic
Stenolaemate Bryozoan Families

ORDER TREPOSTOMATA
  Suborder Esthonioporina
    Dianulitidae, Esthonioporidae, Orbiporidae
  Suborder Halloporina
    Dittoporidae, Halloporidae, Heterotrypidae, Mesotrypidae,
    Monticuliporidae, Trematoporidae
  Suborder Amplexoporina
    Aisenvergiidae, Amplexoporidae, Anisotrypidae, Araxoporidae,
    Atactotoechidae, Crustoporidae, Dyscritellidae, Eridotrypellidae,
    Stenoporidae, Ulrichotrypellidae

ORDER CYCLOSTOMATA*
  Suborder Palaeotubuliporina
    Corynotrypidae, Crownoporidae

ORDER CYSTOPORATA
  Suborder Ceramoporina
    Ceramoporidae
  Suborder Fistuliporina
    Actinotrypidae, Anolotichiidae, Botrylloporidae, Constellariidae,
    Cystodictyonidae, Etherellidae, Fistuliporidae, Goniocladiidae,
    Hexagonellidae, Rhinoporidae, Xenotrypidae

ORDER CRYPTOSTOMATA
  Suborder Ptilodictyina
    Escharoporidae, Intraporidae, Phragmopheridae, Rhinidictyidae,
    Stictoporellidae, Worthenoporidae, Virgatellidae
  Suborder Timanodictyina
    Girtyporidae, Timanodictyidae
  Suborder Rhabdomesina
    Arthrostylidae, Bactroporidae, Hyphasmoporidae, Nikiforovellidae,
    Rhabdomesidae, Rhomboporidae
  Suborder Phylloporinina
    Chainodictyonidae, Chasmatoporidae, Enalloporidae, Phylloporinidae,
    Ralfinidae, Sardesoninidae
  Suborder Fenestellina
    Acanthocladiidae, Fenestellidae, Fenestraliidae, Semicoscinidae,
    Septoporidae

SOURCE: Based on Boardman et al. 1983 and Astrova 1978.
*Tubuliporata in Boardman et al. 1983.

awaits use of the generic-level data set and the representation of each family by multiple genera.

A second cladogram illustrates the effect of changing generic exemplars to ones more typical of each family (figure 8.3). This cladogram is far more congruent with conventional or treatise-based systematics than is figure 8.2. The Fistuliporina are not paraphyletic, and they form a monophyletic group basal to the remaining free-walled (double-walled) clades. The Esthonioporina, in turn, form a basal group to the remaining trepostome–cryptostome clade. Four families of Rhabdomesina form a monophyletic group, which is a sister group to the Fenestrata. The rhabdomesine–fenestrate clade, in turn, is a sister group to the ptilodictyine–trepostome clade. The Ptilodictyina–Timanodictyina are likewise no longer paraphyletic, but they inexplicably include the Ceramoporidae. The Intraporidae are back in the Ptilodictyina, and the Ulrichotrypellidae are back in the Trepostomata. The Amplexoporina have become paraphyletic, but are still linked with the Halloporina in a trepostome clade. With the exception of the aberrant location of the ceramoporines, everything else in this cladogram is very conventional. The Cryptostomata *sensu lato* no longer represent a monophyletic group in figure 8.3 because the ptilodictyines have become linked to the trepostomes. That linkage could be real, however, because some Early Ordovician bryozoans appear to be morphologically intermediate between trepostomes and ptilodictyines, such as *Trepocryptopora* and *Yangotrypa* (Hu and Spjeldnaes 1991). Blake (1983), however, convincingly argued for the monophyly of the Cryptostomata; Blake's argument fits the relationships illustrated in figures 8.1 and 8.2, as well as the cladogram produced by Cuffey and Blake (1991).

When both data sets are analyzed with only the 19 families that first appeared in either the Tremadocian or Arenigian, the two oldest ptilodictyine families are always united with the two oldest phylloporinine families, supporting the concept of cryptostome monophyly (figure 8.8). The oldest trepostomes are always united to the oldest fistuliporines in a monophyletic expletocystid clade, a concept supported by Cuffey (1973) in a phenetic study. Further analysis is required to verify these relationships, but it is clear that some form of stratocladistic algorithm (e.g., Maddison and Maddison 1992) is absolutely essential to separate phylogeny from time-dependent convergent homoplasy. Our results in figure 8.2 produced more apparent homoplasy than those in figure 8.3, but we prefer figures

8.2 and 8.8 for their branching relationships. We acknowledge that figure 8.3 displays more conventional taxonomic arrangements, possibly implying that errors exist in conventional systematics.

Diversity History: Genera Within Suborders

Clade Patterns

Tuckey (1988, 1990a, 1990b) compiled a comprehensive worldwide database of all known Ordovician and Silurian bryozoan species drawn from nearly 500 published sources, and including more than 8,000 records of over 2,100 species; taxonomic assignments used by him followed conventions established by the most recently published synonymies. Horowitz and Pachut (1993) have compiled a similar database for Devonian bryozoans, but excluded separate geographic records for each species. Generic ranges of Carboniferous and Permian bryozoans have been compiled from a much less exhaustive search of the literature, based upon recently published monographs, treatises, and generic compilations. Because a species level database is not yet available for the Carboniferous and Permian, we will present the diversity paths of Paleozoic bryozoans at only the generic level (appendix C). The most useful taxa within which to track diversity patterns are suborders, most of which have been verified as monophyletic clades. Cladistic analysis of conventionally defined orders suggests that they are either polyphyletic or paraphyletic, and not as phylogenetically meaningful as suborders.

Clade diversity diagrams (figure 8.4) represent the generic diversity history of each of the 11 suborders. All but the Timanodictyina originated in the Early Ordovician. Phylogenetic analysis at the family level (figures 8.2 and 8.3) strongly suggests that the Timanodictyina are not a separate clade, but are part of the Ptilodictyina. If the spindle diagrams of the timanodictyines and ptilodictyines were added together, their resultant shape would be very similar to that of the phylloporinines (figure 8.4). The remaining spindle diagrams have only two basic shapes: bottom-heavy, with an Ordovician center of gravity, or rectangular, with a De-

---

Figure 8.4   Clade diversity diagrams of generic diversity with each of the 11 suborders, plotted over the 30 stage/series intervals from Early Ordovician to Late Permian. Three genera of amplexoporines extend into the Norian Stage of the Triassic; the affinities and ages of other Triassic bryozoans await restudy.

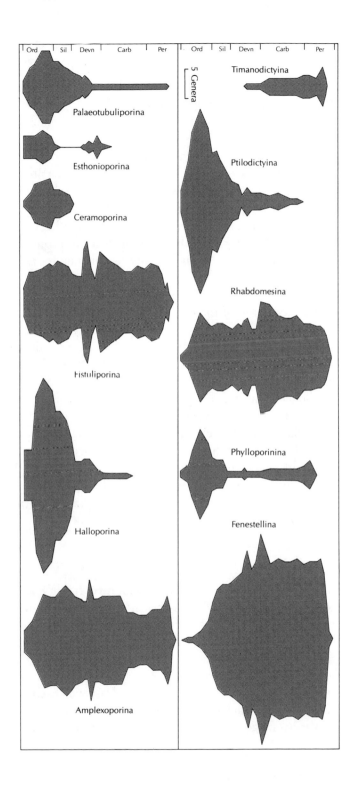

vonian center of gravity. The bottom-heavy clades are the Palaeotubuli-porina, Esthonioporina, Ceramoporina, Halloporina, Ptilodictyina, and Phylloporinina. The rectangular clades are the Fistuliporina, Amplexo-porina, Rhabdomesina, and Fenestellina. The bottom-heavy clades all reached peak generic diversity in the Caradocian or Ashgillian; the rec-tangular clades more or less maintained stable generic diversities but dis-play diversity spikes in both the Givetian and Tournaisian. The rectan-gular clades all terminated in the Late Permian (Triassic survivors other than amplexoporines need reevaluation), and the bottom-heavy clades all declined sharply in diversity from the Late Ordovician through the Late Silurian, with thin spires of residual diversity reaching up to the Carbon-iferous or Permian.

The superimposed diversity patterns of all suborders (figure 8.5) illus-trate the initial expansion of the bottom-heavy clades in the Ordovician and their attainment of a Paleozoic generic maximum in the Caradocian. Devonian through Permian diversity reflects only the contributions of the four rectangular clades. The major change within the Paleozoic is a switch from early co-dominance by both the bottom-heavy and the rectangular clades to a remnant fauna composed predominantly of the rectangular clades. The switch is not a clade replacement phenomenon among bryo-zoans because the rectangular clades are just as generically diverse in the Ordovician and Silurian as they are later in the Paleozoic. Instead, the remnant bottom-heavy clades simply failed to be replaced by other bry-ozoan clades after a series of Ordovician and Silurian extinctions. Late Paleozoic bryozoan faunas never regained the subordinal disparity found before the Devonian.

Evolutionary Faunas

Factor analysis of generic diversity within suborders (table 8.2) over the 30 Paleozoic stages or series reflects the dichotomy between the bottom-heavy (Factor 1) and rectangular (Factor 2) clades (preliminary discussion in Anstey 1990). The same dichotomy applies to standing diversity, ge-neric origins, and generic extinctions. The strong correlation among ori-gin, extinction, and diversity implies maintenance of a dynamic equilib-rium within each of the first two factors, each of which constitutes a within-bryozoa "evolutionary fauna." The parallelism in origin, extinc-tion, and diversity suggests the effects of extrinsic factors, such as niche

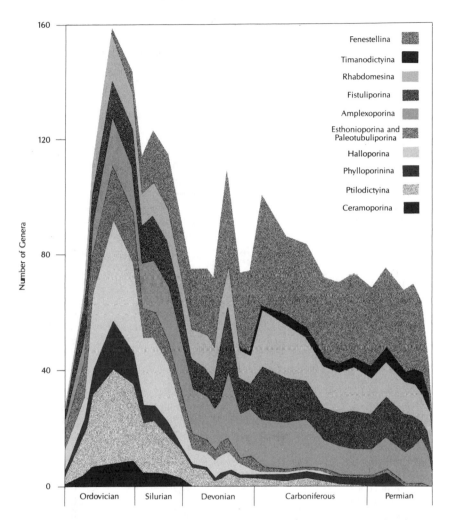

Figure 8.5 Superimposed clade diversity diagram of genera within suborders, plotted over the 30 stage/series divisions of the Ordovician through Permian. Suborders are plotted in the order of factor loadings listed in table 8.2.

availability, that permitted the rapid expansion and decline of the six bottom-heavy clades, and subsequently allowed the four rectangular clades to maintain nearly stable diversities throughout the entire post-Cambrian Paleozoic.

The adaptations of Paleozoic bryozoans to their habitats are not well

Table 8.2

Varimax Rotated Factor Matrices of Generic Diversity, Generic Origins, and Generic Extinctions for Suborders

Suborder	Factor 1	Factor 2
DIVERSITY		
Ceramoporina	0.9628	0.0162
Ptilodictyina	0.9580	0.0858
Phylloporinina	0.9487	0.1452
Halloporina	0.9282	0.1024
Palaeotubuliporina	0.9067	0.1316
Esthonioporina	0.6950	−0.2387
Fenestellina	−0.6445	0.6711
Amplexoporina	0.1592	0.8098
Fistuliporina	0.0880	0.8679
Rhabdomesina	−0.0118	0.8694
Variance	54.6%	25.0%
ORIGINS:		
Ptilodictyina	0.9555	0.1317
Halloporina	0.9211	−0.0200
Ceramoporina	0.8749	0.0112
Phylloporinina	0.8706	0.1243
Palaeotubuliporina	0.8597	0.1397
Esthonioporina	0.0859	0.0175
Amplexoporina	0.3558	0.5968
Rhabdomesina	0.3817	0.7835
Fistuliporina	0.15182	0.7981
Fenestellina	−0.26424	0.8995
Variance	44.1%	20.8%
EXTINCTIONS		
Ptilodictyina	0.9247	−0.1277
Palaeotubuliporina	0.8763	−0.0245
Halloporina	0.8508	−0.1788
Phylloporinina	0.8319	0.0108
Ceramoporina	0.6569	−0.1847
Esthonioporina	0.3786	−0.0511
Fistuliporina	−0.0692	−0.0676
Rhabdomesina	0.3900	0.3287
Amplexoporina	0.0436	0.5172
Fenestellina	−0.2793	0.7530
Variance	38.6%	21.9%

NOTE: Calculations based on 30 Paleozoic stages/series; raw data provided in appendix C.

understood, but available evidence indicates linkages during the Ordovician: Factor 1 clades are linked with major terrigenous epeiric biomes; Factor 2 clades are linked with major carbonate epeiric biomes (Anstey 1986; Tuckey and Anstey 1989, 1992; Tuckey 1990a). Possible adaptive differences in colony design across clades are likewise not understood. One known difference between Early and Late Paleozoic bryozoans is in the depth of the retracted zooid (Anstey 1990), which could reflect an adaptive response to unknown bryozoan predators.

Major Paleozoic Events

Three categories of bryozoan radiations occurred during the Paleozoic (tables 8.3–8.5):

- novelty radiations producing higher taxa and involving the first appearances of specific colonial bauplans;

Table 8.3
Peak Intervals of Radiation and Extinction Among Paleozoic Bryozoans

	Families	Genera
RADIATIONS		
1.	Arenigian (19)	Caradocian (56)
2.	Llanvirnian (9)	Llanvirnian (49)
3.	Caradocian (4)	Tournaisian (42)
4.	Ashgillian (4)	Llandeilian (39)
5.	Givetian (4)	Givetian (33)
6.	Tournaisian (4)	Ashgillian (31)
7.	Ludlovian (3)	Arenigian (31)
EXTINCTIONS		
1.	Dzhulfian (10)	Caradocian (47)
2.	Capitanian (7)	Ashgillian (45)
3.	Pridolian (7)	Givetian (40)
4.	Givetian (5)	Tournaisian (32)
5.	Sakmarian (5)	Capitanian (27)
6.	Ludlovian (4)	Ludlovian (25)
7.	Ashgillian (3)	Pridolian (24)

NOTE: In ranked order (1 = highest, 7 = lowest); number of taxa given in parentheses; all background generic data provided in appendix C.

Table 8.4

Classification of Generic Radiations and Extinctions

RADIATIONS	EXTINCTIONS
Extinction-independent	*Radiation-independent*
1. Llanvirnian (49:o/5:e)	1. Capitanian (27:e/2:o)
2. Llandeilian (39:o/7:e)	2. Ludlovian (25:e/14:o)
3. Arenigian (31:o/3:e)	3. Pridolian (24:e/6:o)
Extinction-balanced	*Radiation-balanced*
1. Caradocian (56:o/47:e)	1. Caradocian (47:e/56:o)
2. Tournaisian (42:o/33:e)	2. Ashgillian (45:e/31:o)
3. Givetian (33:o/40:e)	3. Givetian (40:e/33:o)

NOTE: Number of genera given in parentheses (e = extinct; o = originating); all background data provided in appendix C.

Table 8.5

Higher Taxon Gains and Losses

	RADIATIONS			EXTINCTIONS		
	Suborders	Families	Genera	Suborders	Families	Genera
Arenigian	8	19	31	0	0	2
Llanvirnian	2	9	49	0	0	7
Pridolian	0	2	6	1	7	24
Capitanian	0	0	2	1	7	27
Dzhulfian	0	0	0	3	10	13

NOTE: Radiations and extinctions involving the largest number of higher taxa; all generic background data provided in appendix C.

- extinction-independent generic radiations, producing significant net gains in generic diversity; and
- extinction-linked generic radiations, producing diversity increases balanced by extinction.

The only two Paleozoic intervals of novelty radiation are the Arenigian and Llanvirnian (table 8.5). The major extinction-independent generic radiations took place during the Arenigian, Llanvirnian, and Llandeilian (table 8.4). The major extinction-linked generic radiations took place during the Caradocian, Givetian, and Tournaisian (table 8.4; also, the three largest spikes in figure 8.6). Therefore the interval from the initial appearance of bryozoans in the Tremadocian to the Llanvirnian is marked

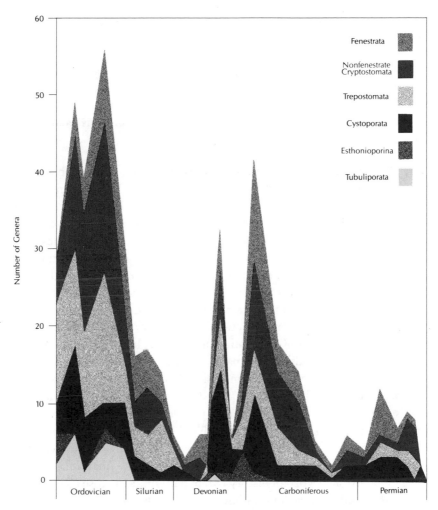

Figure 8.6 Superimposed generic origin diagram of genera within ordinal-rank clades.

by the sustained production of morphological novelties conventionally recognized as higher taxa. The same interval likewise displays heightened generic radiation extending past the novelty event into the Llandeilian and Caradocian, and finally balanced by generic extinctions in the Caradocian. Similar extinction-balanced radiations took place in the Givetian and Tournaisian.

The Paleozoic is marked by three parallel categories of bryozoan extinctions (tables 8.3–8.5):

- higher-taxon extinctions, final appearances of colonial bauplans;
- radiation-independent generic extinctions, producing significant net losses in generic diversity; and
- radiation-linked generic extinctions, producing extinctions balanced by penecontemporaneous radiations.

The three most significant intervals of higher-taxon extinction are the Dzhulfian, Capitanian, and Pridolian (table 8.5). The Capitanian and Dzhulfian represent the two final stages of the Late Permian extinction, and the Pridolian represents the final episode of the Ordovician-Silurian extinctions that had heavily reduced the six bottom-heavy clades. The major radiation-independent generic extinctions took place during the Capitanian, the Pridolian, and the Ludlovian (table 8.4). Thus the Late Silurian extinctions are first pronounced at the generic level in the Ludlovian, and they culminate in family-level extinctions in the Pridolian. Likewise, the Late Permian extinctions are first pronounced at the generic and familial levels in the Capitanian, and they culminate in family-level extinctions in the Dzhulfian. The major radiation-linked generic extinctions took place during the Caradocian, Ashgillian, and Givetian (table 8.4).

The products of both radiation and extinction events are linked to levels of generic diversity and, by proxy, to levels of species diversity. Of the higher-taxon or novelty events (table 8.5), the largest radiation and the two largest extinctions took place during the intervals characterized by Paleozoic minima in generic and species diversity, either at the beginning of Paleozoic bryozoan history (Arenigian) or at its close (Capitanian and Dzhulfian). The generic radiations and extinctions resulting in net changes in generic diversity (i.e., independent events, not balanced ones) took place either at the same time or trailing a novelty radiation, and either at the same time or just preceding a higher-taxon extinction (table 8.4). These net change events all took place in the context of identical or slightly higher generic diversity than that which existed during the higher taxon events.

Last, the episodes of balanced radiation-extinction all took place during episodes of maximal generic and species diversity. The Givetian, for example, currently holds the record for the peak species diversity in the

Paleozoic, with 726 recorded bryozoan species (Horowitz and Pachut 1993). The Early Ashgillian (Tuckey and Anstey 1992) holds the record of highest species diversity of any interval in the Ordovician or Silurian. Peak episodes of species diversity and generic diversity are correlated, representing times of maximum turnover in both species and genera with little or no change in familial or subordinal diversity. The average higher-taxon event took place in the context of an average global diversity of 52 genera, the average net change event in the context of 78 genera, and the average balanced event in the context of 128 genera (tables 8.4, 8.5). Thus there are minimally six categories of diversity flux episodes, and they are correlated with three different levels of species diversity.

Were these diversity flux episodes driven by changes in the number of species, reflecting either speciation rates or species extinction rates? The three balanced radiation-extinction events display a strong correlation between speciation and species extinction. The net change extinctions display generic losses in the Ludlovian and Capitanian, preceding familial and subordinal losses in the Pridolian and Dzhulfian. Therefore the higher taxon extinctions resulted from a collapsing taxonomic hierarchy caused by unreplaced species losses over an extended period. The reverse geometry, however, prevails during the Ordovician radiation, with higher-taxon originations concentrated in the Arenigian and Llanvirnian, followed by heightened generic net increases extending through the Llandeilian. We are thus led to conclude that higher taxon origins did not result from the construction of a hierarchy built by speciation. Instead, the higher taxon origins seem to have set up the potential for heightened speciation in subsequent epochs.

Cladogenesis and Apomorphogenesis

In order to estimate the minimum age of each branching node in our cladograms, as well as the ages of the internode apomorphies, we have made the following assumptions, which utilize stratigraphic order and by which we can convert the cladogram into a stratocladistic tree:

If two taxa, A and B, diverge in the cladogram at node 1, and if A appeared in the fossil record earlier than B, the node 1 is minimally the same age as the first appearance of B. Taxon A predates the node. The minimum age of the synapomorphies plotted below the node, but shared by A and B, is estimated by the age of node 1.

If a third taxon C is joined to the A-B branch at node 2, and if C is older than A, then the first appearance of A in the fossil record minimally estimates the age of node 2. Taxon C predates the node, but synapomorphies shared by the clade A-B-C, plotted below node 2, are minimally estimated by the age of node 2.

If a fourth taxon D is joined to node 3, then the age of node 3 is minimally estimated by the age of C. D, therefore, should be older than node 3. In some cases, however, the first appearance of a taxon, such as D, may be younger than the node to which it is joined. Therefore node 4, which underlies node 3, cannot be estimated by the first appearance of D in the fossil record; it must instead be at least as old as node 3.

Overall, the rule of superposition must be maintained within the cladogram: all lower nodes must be older than higher nodes.

Applying this logic to figure 8.2, the basal node of the Fenestellina must be minimally Llanvirnian in age, because both the Fenestellidae and the unilaminate Arthrostylidae first appeared in the Llanvirnian. The four higher fenestelline nodes are successively Ashgillian (first appearance of Semicoscinidae), Wenlockian (as opposed to Tournaisian first appearance of Septoporidae because the next higher node is Wenlockian), Wenlockian (first appearance of Acanthocladiidae), and Famennian (first appearance of Fenestraliidae). Note that only the Septoporidae are chronologically out of order, and this discrepancy indicates either missing families that evolved between the Wenlockian and Tournaisian or missing character states that would move this branch to a higher node.

Using the same logic, the next 16 basal nodes, from the Phylloporinina down to the Ceramoporina, are all Arenigian in age. The highest of the Arenigian nodes is controlled by the first appearance of the Enalloporidae in the Arenigian; the node immediately below it is controlled by the first appearance of the radial Arthrostylidae in the Arenigian. The next six nodes down, therefore, are minimally Arenigian by the rule of superposition. Because the Anolotichiidae first appeared in the Tremadocian, the six basal nodes of the cladogram are at least Tremadocian in age. New discoveries can increase the age of the first appearance of a taxon, but they cannot decrease its age. Therefore the Fenestellina are at least Llanvirnian in age, or older. Likewise, the Phylloporinina, Rhabdomesina, Ptilodictyina, Amplexoporina, Halloporina, and Ceramoporina are minimally Arenigian in age, and the Fistuliporina, Esthonioporina, and Palaeotubuliporina are minimally Tremadocian in age. Analysis of figure

8.3 does not change any of these age relationships; they are robust beyond the variances in cladistic results.

Many terminal taxa in both cladograms are stratigraphically much younger than the nodes to which they are attached. At least three ad hoc hypotheses may be applied to these discordances. As a first hypothesis, the first appearance of a taxon may be too young because sampling has not yet been adequate to detect it in older strata. As a second hypothesis, undiscovered older intermediates may connect the taxon to its node (recovery of some new families in the fossil record might be achievable through reanalysis of known genera; others may depend upon better recovery of fossils from Tremadocian and Arenigian strata). As a third hypothesis, the branch itself may be placed too low in the cladogram because of undetected character state losses or reversals (bootstrapping techniques indicate that most of the variable nodes are not fixed at a significant level of probability).

The first two hypotheses assume that the cladogram is right, and that stratigraphic sampling is incomplete. The third assumes completeness of stratigraphic sampling, but an insufficiency of character states to place the taxon in its correct cladistic position. In reality, incompleteness exists in both stratigraphic sampling and in character state recovery.

Assuming that gaps of two or more stages (or series) that occur between nodes or between nodes and terminal taxa are stratocladistic discordances, then figure 8.2 has a total of 246 missing stages, or an average of 2.1 per branch segment. Figure 8.3 has 238 missing stages, or an average of 2.0 per branch segment. Therefore, figure 8.3 has a slightly better fit to the available chronostratigraphic data than has figure 8.2; it also has a higher consistency index (0.247 versus 0.232) and a lower homoplasy index (0.755 versus 0.768). Figure 8.2 requires a total of 256 stages (average branch segment duration = 2.18 stages) summed over all of its branches, whereas figure 8.3 requires only 246 (average = 2.07), giving it greater stratigraphic parsimony. Several lines of evidence, therefore, suggest that figure 8.3 might be better than figure 8.2 as an overall representation of Paleozoic bryozoan phylogeny. The differences, however, are not large, and for the sake of this paper both cladograms will be retained as working hypotheses.

Using chronostratigraphic first appearance records, the timing of family-level cladogenetic events can be minimally estimated from both figures 8.2 and 8.3, which display considerable agreement (table 8.6). Most of

Table 8.6
Numbers and Rates of Familial Branching Nodes

Stages/Series	Nodes (number of)		Nodes per Million Years		Nodes per Generic Origin	
	8.2*	8.3	8.2	8.3	8.2	8.3
Tremadocian	6	3	0.35	0.18	2.00	1.00
Arenigian	26	25	1.53	1.47	0.93	0.89
Llanvirnian	9	11	1.20	1.47	0.18	0.22
Llandeilian	1	1	0.21	0.21	0.03	0.03
Caradocian	4	3	0.19	0.14	0.07	0.05
Ashgillian	1	2	0.25	0.50	0.03	0.06
Llandoverian	0	2	0.00	0.22	0.00	0.13
Wenlockian	2	2	0.33	0.33	0.12	0.12
Ludlovian	2	2	0.15	0.15	0.14	0.14
Pridolian	1	3	0.46	1.37	0.17	0.50
Early Devon.	0	0	0.00	0.00	0.00	0.00
Givetian	3	0	0.87	0.00	0.09	0.00
Frasnian	0	0	0.00	0.00	0.00	0.00
Famennian	1	1	0.22	0.22	0.06	0.06
Tournaisian	3	3	0.23	0.23	0.08	0.08
Visean	0	0	0.00	0.00	0.00	0.00
Serpukhovian	0	0	0.00	0.00	0.00	0.00
Bashkirian	0	1	0.00	0.09	0.00	0.20

*Based on figures 8.2 and 8.3.

the 59 nodes (35 or 36) in both cladograms fall within the Arenigian-Llanvirnian interval. More than two-thirds of all familial cladogenesis in the Paleozoic took place within the first 30 million years of the phylum's existence. Familial branching was essentially complete by the Tournaisian, so the final one-third of the branching events took place fairly uniformly over an interval of 120 million years. No familial cladogenesis took place during the final 100 million years of the Paleozoic.

Rates of cladogenesis (table 8.6), based on both cladograms, also identify the Arenigian and Llanvirnian as the two intervals of maximal family branching per million years and per generic origin as well. Tremadocian rates are high per generic origin (given only three known genera) but are only moderate in branchings per million years. Figure 8.3 suggests high rates of branching in the Pridolian as well, but figure 8.2 does not.

The first three stages of the Ordovician produced more than half (56 or 57%) of all Paleozoic familial apomorphies. Levels of apomorphy production (table 8.7) were highest during both the Arenigian and Llanvirnian, in both cladograms. Rates of apomorphy production were maximal as well during the Arenigian and Llanvirnian, and also reached high levels during the Pridolian (table 8.7).

All available metrics based upon all available phylogenetic inferences identify the Arenigian as the peak episode in familial evolution during the Paleozoic. Nevertheless, the Tremadocian-Arenigian has only 60 species known from 33 genera (table 8.8), the lowest number of any stage in the Paleozoic. Of the 33 genera, more than two-thirds (23) are known in this time interval from only one species each. Therefore, barring some unusual bias in stratigraphic sampling, the Tremadocian-Arenigian interval must have had lower levels of speciation than any subsequent Paleozoic stage.

On the basis of information available from the species record, Early Ordovician speciation events must have produced significantly greater

Table 8.7
Numbers and Rates of Familial Apomorphies

Stages/Series	Number		Per Million Years		Per Generic Origin	
	8.2*	8.3	8.3	8.3	8.2	8.3
Tremadocian	26	12	1.5	0.7	8.7	4.0
Arenigian	131	145	7.7	8.5	4.7	5.2
Llanvirnian	51	62	6.8	8.2	1.0	1.3
Llandeilian	8	5	1.8	1.1	0.2	0.1
Caradocian	22	16	1.1	0.8	0.4	0.3
Ashgillian	19	25	4.6	6.0	0.6	0.8
Llandoverian	2	6	0.2	0.6	0.1	0.3
Wenlockian	8	8	1.2	1.2	0.4	0.4
Ludlovian	13	12	1.0	0.9	1.0	0.9
Pridolian	12	19	5.5	8.7	2.0	3.2
Early Devon.	0	0	0.0	0.0	0.0	0.0
Eifelian	7	4	1.3	0.7	0.3	0.2
Givetian	15	11	4.4	3.2	0.5	0.3
Frasnian	0	0	0.0	0.0	0.0	0.0
Famennian	2	2	0.4	0.4	0.2	0.2
Tournaisian	33	23	2.6	1.8	0.7	0.5
Visean	3	4	0.2	0.3	0.2	0.3
Serpukhovian	4	6	0.4	0.6	0.3	0.5
Late Carbon.	9	14	0.3	0.5	0.7	1.1
Early Permian	6	8	0.2	0.3	0.3	0.4

*Based on figures 8.2 and 8.3.

Table 8.8

Bryozoan Genera and Species of the Early Ordovician

TREMADOCIAN STAGE
Orbiporidae: *Hubeipora*, 2 unnamed species; *Yichangopora*, 4 unnamed species
Anolotichiidae: *Profistulipora arctica, retrusa, unapensis*

ARENIGIAN STAGE
Corynotrypidae: *Wolinella baltica*
Crownoporidae: *Sagenella*, 1 unnamed species
Dianulitidae: *Dianulites fastigiatus, glauconiticus, Hexaporites janischewskyi, multimesoporina, petropolitana, utahensis*; *Hexaporites*, 1 unnamed species; *Revalotrypa gibbosa*
Esthonioporidae: *Esthoniopora communis, curvata, lessnikovae*
Orbiporidae: *Orbipora acanthopora, solida*, 1 unnamed species; *Yichangopora mui*
Anolotichiidae: *Lamtshinopora hirsuta*
Xenotrypidae: *Xenotrypa primaeva*
Dittoporidae: *Dittopora annulata, clavaeformis, ramosa, sokolovi*; *Hemiphragma rotundatum*
Halloporidae: *Hallopora*, 1 unnamed species; *Diplotrypa bicornis, petropolitanus*
Heterotrypidae: *Annunziopora*, 1 unnamed species
Monticuliporidae: *Monticulipora*, 1 unnamed species
Trematoporidae: *Batostoma*, 1 unnamed species; *Eridotrypa*, 1 unnamed species; *Lammotopora*, 1 unnamed species; *Nicholsonella arborea, huoi, papillaris, rotundicellularis*, 4 unnamed species
Amplexoporidae: *Amplexopora*, 1 unnamed species; *Anaphragma vetustum*; *Monotrypa helenae*
Atactotoechidae: *Cyphotrypa*, 1 unnamed species
Rhinidictyidae: *Phyllodictya crystallaria*; *Prophyllodictya intermedia*
Stictoporellidae: *Stictoporellina gracilis*
Uncertain ptilodictyine family: *Trepocryptopora dichotoma, flabellaris*; *Yangotrypa*, 1 unnamed species
Arthrostylidae: *Arthroclema* cf. *armatum*
Enalloporidae: *Alwynopora orodamnus*

amounts of morphological evolution than speciation at any later time, i.e., quantum evolution in the sense of Simpson (1944). The morphologies recognized as higher taxa of bryozoans were not built up through a gradual accumulation of species differences but appear to have diverged very rapidly in the initial radiation of the phylum. By the Llanvirnian, all the higher taxa (orders and suborders) had come into existence. Rates of speciation then increased in the Llandeilian and Caradocian. The processes producing the major branching events and familial apomorphies, therefore, apparently were not driven by speciation, and likewise could not have resulted from species selection or species sorting.

Reality of Bryozoan Higher Taxa

Two reviewers of our work have expressed significant concerns regarding the reality of bryozoan higher taxa. One suggested that higher taxa are somewhat arbitrary categories that are largely based on the morphological novelties that arose during the Early Ordovician. Another suggested that the critical character states of Early Ordovician bryozoans subconsciously reflect the dominant taxa of the later Paleozoic; rudimentary states that might be better developed in later taxa might have been given undue weight.

Early Ordovician bryozoans were known from only 8 genera in 1911, 14 by 1953, 28 by 1970, and all 33 listed in table 8.8 by 1991. The taxonomy of bryozoan higher categories was essentially developed prior to an adequate knowledge of early bryozoans. The phenomenon that took place was the recognition that early bryozoans belonged to the same categories as those observed in younger strata. Therefore the second expressed concern—that of possible shoehorning of early forms into later categories—needs to be addressed.

Because so many derived higher taxa seem to originate in the Arenigian, one must ask if the presumed genera from derived clades are correctly assigned. Of the 33 genera listed in table 8.8, two are fixed-walled and consequently belong to the most primitive grade known. An additional seven genera are esthonioporines, representing the most primitive grade among all free-walled bryozoans. Three additional genera are simple fistuliporines, the second most primitive grade. Therefore more than a third of the fauna represents primitive groups and cannot be suspect of shoehorning. Of the trepostomes present, ten genera have simple encrusting or hemispherical growth forms. Only eleven genera represent the more derived clades. All but three of these have been described or reevaluated in modern papers published since 1970. These more modern descriptions include high quality photographs, and they provide enough descriptive information to satisfy skeptics. A good example is *Alwynopora* (Taylor and Curry 1985), the oldest known phylloporinine fenestrate bryozoan. The illustrations include eleven scanning electron micrographs of both external and internal morphology, supplemented by three line drawings and four pages of text description. There is little doubt that *Alwynopora* is the oldest known member of a highly derived clade.

The other question raised also asks if the higher taxa of bryozoans are truly clades—that is, evolutionarily real higher taxa that arose through branching processes. Our response, illustrated in figures 8.2 and 8.3, is that the Paleozoic stenolaemate orders and suborders are indeed clades or paraclades (as discussed in the foregoing text) and that these groups did arise through the progressive acquisition of derived characters through a process of evolutionary branching. The higher taxa of bryozoans, in our view, are not phylogenetically unrelated groups of novel morphologies; they are cladistically defined.

Speciation History of Paleozoic Bryozoans

The history of species diversity has been plotted through the Ordovician and Silurian by Tuckey and Anstey (1992, their figure 3), and through the Devonian by Horowitz and Pachut (1993, their figure 4). Because species diversity is strongly correlated with generic diversity in the Ordovician through Devonian, it can be inferred by generic proxy for the entire Paleozoic. (See figures 8.4–8.7.)

One goal of this paper is to assess changes in morphological distances achieved by speciation events at different times in the history of Paleozoic bryozoans. By comparing production of familial branching and apomorphies with the species record, it is clear that the bulk of familial morphological evolution took place either cryptically (in such a way as to leave no record) or in the absence of extensive speciation. The cryptic hypothesis lacks support because bryozoan taxa do appear in cladogram order in the stratigraphic record. Tremadocian bryozoans (table 8.8) are esthonioporines and fistuliporines. The six suborders that come next in the cladogram all appeared in the Arenigian. The fenestellines are the last to appear stratigraphically, and they also occupy the highest cladogram position in figure 8.2. If more derived taxa had been found, for instance, in the Tremadocian, then the cryptic hypothesis would have more support. For the time being, therefore, we have to assume that rich undiscovered bryozoan faunas will not overturn the diversity pattern illustrated in figures 8.4–8.7. If one assumes that the species retrieved are representative samples of the diversity that actually existed during each stage, then it

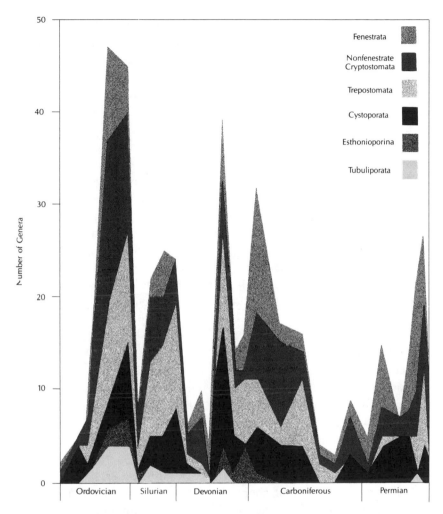

Figure 8.7 Superimposed generic extinction diagram of genera within ordinal-rank clades.

can be concluded that the earliest speciation in bryozoans produced much greater quantum distances than did speciation at all later Paleozoic times.

Generic phylogeny is illustrated, in part, in figures 8.2 and 8.3, because each family is represented by one generic exemplar. The generic character set will, for the most part, add only those characters that are autapomorphic within suborders; it might also resolve some of the conflicts between figures 8.2 and 8.3. In the Tremadocian-Arenigian, the relationship between genera and families is very close, with 33 genera representing 19 families, for an average of 1.8 genera per family. The ratio increases to 2.7 in the Llanvirnian, to 3.8 in the Llandeilian, to an Ordovician maximum of 4.8 in the Caradocian. In the Early Ordovician, therefore, the rates and timing of generic evolution will be very close to that of the family-level cladograms (tables 8.6–8.7).

The Ordovician radiation of bryozoans produced a new phylum, two classes (Stenolaemata and Gymnolaemata), and all the Paleozoic orders and suborders. By the close of the Ordovician, the ongoing radiation had produced 76–80% of all familial branching in the Paleozoic, 69–71% of all Paleozoic familial apomorphies, 62% of all Paleozoic families, and 44% of all Paleozoic genera. The radiation was essentially damped by the Late Ordovician extinction (Tuckey and Anstey 1992), as well as by continuing extinctions through the Silurian. The net effect was the severe reduction of the bottom-heavy clades in figure 8.4, leaving a residual late Paleozoic fauna made up of the four rectangular clades.

Homoplasy

Homoplasy is a dominant theme in bryozoan evolution. It includes, by definition, three phenomena: the loss, re-evolution, and multiple evolution of derived character states. Losses are reversions to primitive states. Re-evolution is the reappearance of a previously lost derived state. Multiple evolution, including both convergence and parallelism, is the multiple appearance of the same derived state in either closely or distantly related lineages. In figure 8.2, the total of the apomorphies defining each branch includes 29.5% of character state losses (reversions to plesiomorphy), 6.2% of reappearances of derived states within the same lineage (following a loss, and excluding reappearances resulting from losses), and 55.1% multiple appearances of derived states in different lineages (excluding convergences that result from losses or from reappearances). Therefore only 9.2% of the available apomorphies represent totally monotonic anagenesis (unidirectional change toward more derived states). Of the anagenetic steps in the cladogram, 79.4% are concentrated in the Early Ordovician, and homoplasy therefore dominates the remaining Paleozoic.

To determine the pattern of initial branching in Paleozoic bryozoans, no taxa appearing after the Llanvirnian should be included, thus avoiding the effects of time-dependent convergent homoplasy. Using only the 29 families that had appeared by the Llanvirnian, even different data sets yield nearly identical results (figure 8.8). The cryptostomes form a mono-phyletic group (Blake 1983) consisting of ptilodictyines, arthrostylids, and fenestrates. The remaining trepostomes and cystoporates form an expletocystid clade (Cuffey 1973), with the esthonioporines separate from the remaining cystoporate-trepostome clade. The branching pattern is nearly identical to that in figure 8.1, except that in figure 8.1 the exple-tocystids are not united. Adding in the five additional families that had appeared by the Caradocian does not alter the branching pattern. The major change in the branching pattern arises by adding in two families

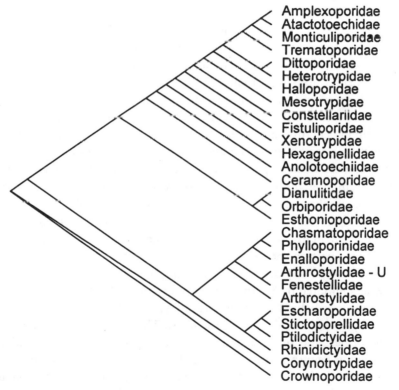

Figure 8.8 Cladogram of the 28 families of Paleozoic stenolaemate bryozoans present in the stratigraphic record by the Llanvirnian. Arthrostylidae-U = uni-laminate arthrostylids.

of rhabdomesines that first appeared in the Ashgillian (Rhabdomesidae and Bactroporidae); these "destroy" the cryptostome clade by becoming a convergent bridge between ptilodictyines and trepostomes, producing a pattern exactly like that in figure 8.3. These two "problem" families are highly convergent between cryptostomes and trepostomes. The Rhomboporidae are even more convergent, but first appear in the Wenlockian.

Therefore the emerging picture of bryozoan evolution is one that puts the establishment of the basic patterns and designs at the very beginning. All subsequent evolution is nothing but variation on the genesis themes. Bryozoan disparity, or the spectrum of colonial construction designs, was complete by the Llanvirnian. By the Ashgillian, convergence became sufficiently strong to mask some of the early branching relationships. Large-scale convergences seem to dominate most post-Caradocian bryozoan evolution.

Some of the major convergences cause considerable difficulty in phylogenetic analysis. One such convergence is the multiple evolution of the bifoliate colony form in ptilodictyines, fistuliporines, and some genera of ceramoporines and trepostomes. The rhabdomesines, timanodictyines, ceramoporines, and some fistuliporines display additional convergences in the multiple evolution of trepostome-like character states. Furthermore, simple encrusting colonies in several suborders converge in form through the loss of derived character states associated with erect colony forms. Some Paleozoic genera and species have convergent character states resembling those found in post-Paleozoic cheilostomes. Several Mesozoic tubuliporate (cyclostome) bryozoans also display convergent Paleozoic character states, which prompted Boardman (1984) to declare that the Permian extinctions were simply an artifact of taxonomic oversight. That question should be resolved through the ongoing restudy of the morphology, phylogeny, and stratigraphic relationships of Triassic bryozoans initiated by Schaffer and Fois-Erickson (1986).

The reasons behind rampant homoplasy in Paleozoic bryozoans are essentially architectural, or constructional. The palaeotubuliporines, esthonioporines, and some fistuliporines and trepostomes first evolved as encrusting or hemispherical colonies. Different styles of erect colony construction evolved as modifications of encrusting colonies. Erect bifoliate colonies developed as vertical sheets from an upfolding and doubling over of the basal lamina. Erect radial colonies developed as stems or trunks cored by a bundle of vertically extended axial zooids, or axial endozone.

Erect unilaminate branching colonies developed in turn from the reduction of the axial endozone and the loss of zooids on one side of the branch.

These four styles of colony construction have all reappeared in Mesozoic and Cenozoic bryozoans as well, in both stenolaemates and gymnolaemates. In Paleozoic bryozoans, the unilaminate style of construction appears to be monophyletic (figures 8.2, 8.3, and 8.8), whereas the other three styles appear in many different parts of the cladogram at the family level and even more pervasively at the generic level. Therefore they represent colony designs that have been "discovered" multiple times by natural selection.

The construction of a colony is inherently an astogenetic process, and the potential for following multiple constructional pathways lies within the norm of reaction of many taxa, or is easily modified through evolutionary change (Anstey 1987). The pervasiveness of intraspecific and intracolonial heterochrony suggests that speciation is not necessary to force colony growth into alternative channels. Alternate patterns of fixed growth may explain the high levels of homoplasy in bryozoan evolution and much of the variability within species, as well as much ecophenotypic plasticity (Pachut 1987, 1989; Pachut, Cuffey, and Anstey 1991). The Bryozoa, and the Class Stenolaemata in particular, comprise organisms whose essential modular unit, the zooid, seems to have undergone little modification, or modifications that we are not aware of. Astogeny consists of the addition of zooids and the changes, through growth, in their positions and geometric arrangements, and the varying development of extrazooidal morphology. Evolutionary change has produced species and higher taxa, but even their recognition and their phylogenetic relationships have been confounded with the relative ease that colonial morphology slips into alternative astogenetic pathways.

The Species Problem: Heritability and Variability

Bryozoans constitute ideal candidates for the study of microevolution and speciation in the fossil record. Species differences within genera are predominantly in biometric characters, and can be tested statistically. Most important, the nonheritable within-colony component of variance can be separated from the potentially heritable among-colony component, and heritabilities can be estimated.

Pachut (1989, table 9) calculated heritabilities of four biometric characters in four species of Ordovician bryozoans. He found that these four characters varied in heritability from 7 to 82%. He also calculated heritabilities in two Devonian bryozoan species (Pachut 1982, table 2, among-colony column) for 21 characters, which ranged from 0 to 99%. Heritabilities, however, cannot be estimated directly in most solitary fossil organisms, and stasis or gradualism cannot be easily separated from ecophenotypic variability. It is not known, for example, if trends in the size or shape of fossil mammals, trilobites, fusulinids, or oysters are nonheritable responses to an unknown environmental trend.

Heritabilities have so far been estimated in ten species of Paleozoic bryozoans (table 8.9) for zooecial characters, particularly the size and spacing of zooecia, which are useful in species-level diagnosis. Heritabilities tend to decrease in onshore and low-diversity populations by an intermonticular average of 15% in Edenian bryozoans (Pachut 1989) and 26% in Maysvillian bryozoans (as calculated from mean squares provided by Key 1987). The loss of heritable variation in more dynamic habitats implies that the colonies responded to physical challenges via an increase in nonheritable plasticity, as a broader realized norm of reaction.

In two of the ten species, *Fistulipora decora* and *Heterotrypa frondosa*, mean heritabilities of both zooecial size and spacing fell below 30% (table 8.9). In such species, there is little hope of recovering microevolutionary variation, because the nonheritable variation, in addition to the normal errors of sampling, biometric measurement, and random error, simply

Table 8.9
Mean Heritabilities of Ten Species of Paleozoic Bryozoans

Species	Zooecial Size or Density	Zooecial Spacing
Nicklesopora renzettiae	71.0%	76.5%
Leptotrypella pellucida	34.9%	79.0%
Peronopora vera	63.0%	38.0%
Parvohallopora nodulosa	56.0%	42.0%
Amplexopora septosa	37.5%	59.0%
Heterotrypa ulrichi	39.5%	18.0%
Homotrypa obliqua	37.2%	16.6%
Parvohallopora ramosa	30.1%	17.3%
Fistulipora decora	25.0%	28.3%
Heterotrypa frondosa	21.5%	18.2%

SOURCE: Data published in Farmer and Rowell 1973; Pachut 1982, 1989; Key 1987.
NOTE: Species listed in rank order from highest to lowest.

overwhelms any weak heritable pattern. In such species, evolutionary variation cannot be detected without very large sample sizes for each population sampled within a measured section.

The pattern of heritability levels in table 8.9 is broadly phylogenetic. One of the lowest heritabilities is in a primitive clade, the fistuliporines, represented by the genus *Fistulipora*. This genus has more than 300 named species (Horowitz 1970), many of which are very likely ecophenotypic variants rather than species. The highest observed mean heritability belongs to *Nicklesopora,* a member of a highly derived clade, the rhabdomesines (a sister group to the fenestrates). The intermediate heritabilities all belong to trepostomes, which are phylogenetically intermediate as well.

Low levels of heritability across all populations of a species imply a locked-in evolutionary stasis in the observed characters, even though total observed variability may appear to be highly dynamic or oscillatory within stratigraphic sections. Such species have acquired a strategy of nonheritable response to habitat changes, and heritable genetic change is adaptively unnecessary. Fistuliporids, for example, seem to fit this pattern. Observations indicate that fistuliporid astogeny (Perry and Hattin 1958) is highly dynamic but lacks fixed patterns of colony growth. Fistuliporids are stratigraphically abundant, but their plasticity and chaotic taxonomy (Horowitz 1970) has generally prevented them from serving as useful index fossils.

Four Edenian bryozoan species (Pachut 1987) extend stratigraphically through the entire Kope Formation, over a stratigraphic interval of 60 meters, with an estimated duration of 7.3 million years (Anstey and Rabbio 1989). These species have mean biometric heritabilities of 51, 49, 48, and 29% (table 8.9). In three of them, close to half of the observed variation is heritable; stratigraphic variation in the heritable component reflects microevolution. Preliminary data indicate that all of these species display concurrent stasis in autozooidal characters during the Edenian, even in populations differing significantly in heritability.

A simple strategy for detecting microevolutionary changes in heritable characters is to calculate colony means and then to estimate population mean and variance as the mean and variance of the colony means. This technique, advocated by Anstey and Perry (1970), measures only the among-colony variation, strongly reduces the nonheritable variation, and provides other statistical benefits as well. Such data were provided by

Prezbindowski and Anstey (1978) for two Devonian species, by Pachut (1987) for four Ordovician species, and by Brown and Daly (1985a) for three species of *Parvohallopora* sampled over the entire thickness of the Dillsboro Formation in southeastern Indiana (approximately 77 meters, beginning at the top of the Kope Formation).

Brown and Daly's (1985a) study is based on six biometric characters, and it is the only published analysis of evolutionary tempo in Paleozoic bryozoans. Their technique, plotting the means of the colony means for each species from each sampling interval, automatically reduces the nonheritable component of variation. Their analysis includes three species, *P. ramosa, P. subplana,* and *P. subnodosa.* They consider *P. ramosa* to be a senior synonym of *P. nodulosa* (Brown and Daly 1985b); Anstey and Perry (1973) considered *P. subplana* to be a junior synonym of *P. nodulosa.* Heritabilities of zooecial characters in this lineage range from 17 to 56% (table 8.9). It is conceivable, therefore, and supported by published synonymies, that *P. ramosa, nodulosa,* and *subplana* all represent a single species with a large component of nonheritable variation.

Some of the stratigraphic intervals sampled by Brown and Daly show possible microevolutionary gradualism. The strongest trend is an increase in mesozooecial density in *P. ramosa* over a 29 meter interval. But this trend is based on only six data points within the interval, as *P. ramosa* is missing in 14 of the 20 sampling zones. Two additional points in their study from a nearby correlated section would strongly reverse the upper part of this trend. Segments of local measured sections show well-sampled sustained trends in other characters over 7 to 8 meters of section. The bulk of the Brown and Daly data, however, shows fluctuating variation around the species means for the six measured characters. In our view, these populations demonstrate no sustained gradual trends consistent from one section to another, but they do exhibit a very dynamic form of oscillatory stasis over the 77 meters sampled.

The heritable component of variance (i.e., additive genetic variance) has not yet been examined over any long stratigraphic intervals. This component of variance (which is the numerator of the heritability ratio) could well be independently related to stasis. For example, species displaying stasis might have been maintaining a high level of population heterozygosity (Palmer 1986), with gene flow preventing shifts in genetic equilibrium. On the other hand, a microevolutionary trend or fluctuation could possibly be related to a short-term loss in genetic variability within populations. Even dynamic stasis could consist of short-lived genetic bottlenecks generating fluctuations in population means. Plotting normalized heritable (i.e., additive genetic) variance (as opposed to heritability) stratigraphically against changes in population means could track microevolutionary changes in the context of population genetics (Pachut 1987).

Clinal variations are biometrically documented and statistically verified in eight species of Paleozoic bryozoans (Cuffey 1967; Pachut 1987, 1989; Key 1987; Pachut and Cuffey 1991). These clines existed from more onshore to more offshore populations, or from shallower to deeper water, or from assemblages of lower diversity to those of higher diversity. Several patterns are common to all of these clines: increased plasticity and loss of heritability onshore (accompanied by colonial paedomorphosis) and decreased plasticity, gain in heritability, and colonial peramorphosis offshore.

Stratigraphic fluctuations, such as those plotted or described by Brown and Daly (1985a), Pachut and Cuffey (1991), and Key (1987), reflect habitat clines developed over time in local sections. Short-duration trends in characters, particularly those sustained vertically over 10–20 meters of epeiric section, are clinal shifts rather than microevolution within a bounded epeiric habitat. Even longer-term fluctuations within Cincinnatian lineages, for example, may reflect long-term cycles of sea level changes (Anstey, Rabbio, and Tuckey 1987) and may fall within the normal range of responses of a long-lived species displaying oscillatory stasis.

All of the bryozoan species that have been sampled over lengthy stratigraphic intervals were exceptionally long-lived. All have estimated durations exceeding 7 million years (Anstey and Rabbio 1989) and must have been exceptionally abundant as well. In some fair-weather beds, "thickets" of *Parvohallopora* produced hundreds of colony fragments per square meter of sea floor. Contemporaneous populations were distributed over distances exceeding 600 km of epeiric seaway (Anstey 1986); dense, widespread populations must have been maintained for millions of years. Therefore, based on population size alone, such species would be expected to display stasis (Lande 1985).

By no means are all Paleozoic bryozoan species long-lived or highly variable. Many "bursts" of nomenclature exist within limited stratigraphic intervals. For example, the uppermost Richmondian strata of Indiana and Ohio include the apparently rapid appearance of multiple new species of *Homotrypa, Peronopora*, and *Constellaria* (Utgaard and Perry

1964; Cutler 1968). The uppermost Shermanian strata in New York state include an analogous proliferation of *Prasopora* species (Ross 1967). Similar examples exist in Silurian, Devonian, and Carboniferous strata, usually associated with stratigraphic intervals that are particularly bryozoan-rich. These examples probably do not suggest punctuated or condensed-interval morphospeciation, because the simultaneous appearance of multiple new species also suggests the probability of allogenic introduction of taxa from elsewhere, usually connected with a biome-scale facies change (Anstey 1986). Terrigenous units within carbonate sequences contain some of the richest local bryozoan faunas, such as the Decorah, Waldron, and Rochester Shales.

Early Large Effects, Subsequent Homoplasy

The phylogeny of Paleozoic bryozoans can be determined using available morphological data, but not without some uncertainty. The majority of Paleozoic suborders represent clades, but their initial branching relationships in the Early Ordovician are in part still unresolved. Taxa higher than suborders await phylogenetic definition.

The diversity history of Paleozoic bryozoans involves two major groups of clades. One group is bottom-heavy in time, with peak diversity in the Ordovician, and extinction or major decline in the Silurian. The other group is rectangular, with nearly steady diversities throughout the Paleozoic, and extinction or major decline in the Late Permian.

Early Ordovician speciation apparently produced most of the clade branching and family-level derived character states in the entire Paleozoic. The initial radiation of the phylum does not seem to have been driven by ordinary speciation processes. The morphological distances achieved by early speciation events, if the species record is not unusually biased, must have been much greater than at any subsequent time. The morphological distances thus represent quantum evolution.

The subsequent record of Paleozoic bryozoan evolution is one of pervasive homoplasy, or continuous reiterations of the morphological states that appeared at the outset. Homoplasy is attributed to the apparent intrinsic ease of alteration in the constructional aspects of astogeny. Abundant and long-surviving Paleozoic species display dynamic stasis. In the Paleozoic clades of this phylum, speciation was not a constant phenomenon over time.

Appendix A

Characters Used in Cladistic Analysis of Bryozoan Suborders

Note: Figure 8.1 is based on these characters; character states are additively coded, with 0 as primitive.

1. ZOARIAL FORM: encrusting threads (0); encrusting to massive (1); elevated as fronds or branches (2).

2. ZOARIAL FORM: nonelevated or encrusting (0); cylindrical branches (1); fenestrate lattices (2).

3. ZOARIAL FORM: nonelevated or encrusting (0); cylindrical branches (1); flattened branches or fronds (2).

4. BUDDING: one-dimensional (0); two-dimensional (1); three-dimensional (2).

5. WALL BOUNDARIES: separate (fixed-walled) (0); fused (free-walled) (1).

6. ENDOZONE-EXOZONE: not distinct (0); somewhat distinct (1); markedly distinct (2).

7. CHANGE BETWEEN ENDOZONE AND EXOZONE: none (0); very gradual (1); moderately gradual (2); abrupt (3).

8. LENGTH OF ENDOZONAL PORTION OF ZOOECIUM: not distinguishable (0); short (1); intermediate (2); long (3).

9. WALLS OF UNIFORM THICKNESS: yes (0); no (1).

10. MESOZOOECIA: NO (0); YES (1).

11. METAPORES: no (0); yes (1).

12. EXILAZOOECIA: no (0); yes (1).

13. MONTICULES/MACULAE: none or rare (0); present, not rare (1); radial to stellate (2).

14. MONTICULES/MACULAE: none or rare (0); present, not rare (1); ringed by large polymorphs (2).

15. WALL MICROGRANULAR: no (0); yes (1).

16. WALL LAMINATED: no (0); yes (1).

17. INTERZOOECIAL PORES: no (0); yes (1).

18. LUNARIA: no (0); yes (1).

19. ACANTHOSTYLES: none (0); sparse (1); abundant (2).

20. COMPLETE DIAPHRAGMS: none (0); sparse (1); abundant (2).

21. HEMIPHRAGMS: no (0); yes (1).

22. HEMISEPTA: no (0); yes (1).

23. OVICELLS: no (0); yes (1).

24. SKELETAL GROWTH CYCLES (CESSATION SURFACES): no (0); yes (1).

25. BUDDING: nonhexagonal pattern (0); moderately hexagonal (1); linear-rhombic pattern (2).

26. STENOSTYLES ("CAPILLARIES") OR MURAL STYLES: no (0); yes (1).

Appendix B

Characters Used for Cladistic Analysis of Families of Paleozoic Stenolaemate Bryozoans

Note: Figures 8.2 and 8.3 are based on these characters; characters 1, 2, 21, and 23 were omitted in both analyses.

ZOARIAL FORM:

1. Encrusting, uniserial threads (0); encrusting, multiserial thin sheets (1); multiserial colonies elevated above the substratum, including multilaminar, hemispherical, cylindrical, hollow ramose, or sheetlike (unilaminate, bilaminate, and radially frondescent) forms (2); multiserial, multilaminar, or nonlaminar amorphous elevated colonies ("massive") (3).

2. Encrusting or elevated (meaning vertical growth exceeding that of a thin sheet) (0); hemispherical (1); cylindrical or flattened cylindrical stem (2); cylindrical, dichotomously branching stems (3).

3. Encrusting or elevated, not hollow ramose (0); hollow ramose (1).

4. Encrusting or elevated, nonbilaminate (0); explanate bilaminate fronds (1); flattened branches with mesothecae (2); flattened branches with mesothecae and nonzooidal margins (3).

5. Encrusting or elevated, nonunilaminate (0); erect unilaminate stem, bearing two or more vertical rows of zooids (1); multiple erect unilaminate free branches (2); branches anastomosing to form a cribrate colony (3); branches united by dissepiments bearing up to two rows of zooids (4); branches united by nonzooidal dissepiments (5).

6. Nonunilaminate (0); branches with more than two vertical rows of zooids (1); branches with only two vertical rows of zooids (2).

7. Nonfenestrate or noncribrate (0); fenestrules rhombic or polygonal (1); fenestrules oval or rounded (2); elongate (3).

8. No endozone-exozone distinction present (0); endozone confined to recumbent portions of zooecia, or shorter than one zooecial diameter (1); endozone of intermediate length, extending 1–3 zooecial diameters (2); long endozone, extending more than 3 zooecial diameters (3).
9. No endozone-exozone transition in wall thickness (0); transition very gradual (1); moderate (as in most trepostomes) (2); abrupt (as in most cryptostomes) (3).
10. No change in zooecial orientation at endozone-exozone boundary, or boundary absent (0); reorientation angle low, 0–30 degrees (1); moderate, 30–60 degrees (2); high, 60–90 degrees (3).

BUDDING AND ZOOECIAL SHAPE:

11. Uniserial colonies (0); multiserial colonies, zooids hexagonally packed or lacking strong alignments (1); rhombic or rhombic-linear alignments (2); longitudinal rows of zooids (3); strongly linear alignment, with longitudinal range boundaries (4).
12. Colonies lacking localized budding centers within branches or stems (0); budding localized along a linear endozonal axis or plate (1); originating from an axial zooid (2) or a localized zooid cluster (3).
13. Zooecia nearly perpendicular to zoarial surface (0); zooecia oblique to zoarial surface (1).
14. Uniserial colonies (0); polygonal apertures in multiserial colonies (1); rounded apertures in multiserial colonies (2).
15. Basal portion of zooecia same diameter or smaller than distal portions (0); zooecia inflated or hemispherical in basal portions (1).
16. Length of autozooecia (including portions below living chamber) short, 1–2 zooecial diameters (0); intermediate, 2–4 diameters (1); long, 4 or more diameters (2).

ZOOECIAL WALLS:

17. Walls entirely microgranular (0); microgranular core and laminated outer portions of walls (1); walls between zooids entirely laminated (2).
18. Walls lacking laminated portions (0); geometry of wall laminations: longitudinal (1); V-shaped (2); U-shaped (3); transverse across broad regions (4).
19. Walls of uniform or uniformly increasing thickness (0); walls unevenly thickened but not distinctly beaded (1); walls distinctly beaded (2).
20. Interzooecial walls solid (0); perforated by rare or few mural pores (1); regularly perforated by mural pores (2).

21. Mesotheca absent (0); mesotheca with continuous microgranular zone (1); with discontinuous microgranular zone (2); completely laminated (3).
22. Mesotheca absent (0); median tubules absent or rare in mesotheca (1); median tubules common in mesotheca (2).
23. Laminated stereom absent in zoarium, or absent in intermonticular exozone (0); present in intermonticular exozone (1); thickly developed in exozone (2); stereom laminae crinkled in exozone (3).
24. Lunarium absent (0); lunarium granular-prismatic or variable (1); hyaline or laminated (2).
25. Peristomial ridges absent (0); present (1).
26. Longitudinal ridges absent between rows of zooids (0); dividing rows of zooids (1); extending into zoarial superstructure (2).
27. Extensive nonzooidal surfaces absent (0); smooth (1); marked by longitudinal ridges (2).

POLYMORPHS/EXTRAZOOIDAL STRUCTURES:

28. Cystopores (vesicles) absent (0); present but sparse (1); abundant (2).
29. Cystopores absent (0); present in exozone or throughout zoarium (1); concentrated in exozone (2).
30. Cystopores absent (0); large and irregularly shaped (1); small, boxlike or blisterlike (2).
31. Ovicells absent (0); present (1).
32. Mesozooecia absent (0); sparse (1); abundant (2).
33. Mesozooecia absent (0); open at surface (1); closed over at surface (2).
34. Metapores absent (0); sparse (1); abundant (2).
35. Exilazooecia absent or very rare (0); rare to sparse (1); numerous (2).
36. Autozooecia contiguous, on all sides in multiserial colonies or along a chain of zooids in uniserial colonies (0); partially isolated by small polymorphs or stereom in multiserial colonies (1); completely isolated (2).
37. Monticules absent or rare (0); flat or depressed (1); elevated (2).
38. Monticules absent or lack centers formed by stereom (0); have centers formed by stereom (1).
39. Monticules absent or lack centers formed by clusters of small polymorphs or vesicles (0); have centers of clustered small polymorphs or vesicles (1).
40. Monticules absent or lack megazooecia (0); megazooecia present in monticules (1); megazooecia surround central area formed by stereom or by clustered small polymorphs or vesicles.

41. Monticules absent or rare (0); circular or oval (1); radial to stellate (2).
42. Monticules absent (0); present, but irregularly spaced (1); regularly spaced (2).

STYLES/STYLETS:

43. Acanthostyles absent or rare (0); sparse (1); abundant (2).
44. Acanthostyles absent (0); unimodal size distribution in zoarium (1); simultaneously present as two or more distinct size classes (2).
45. Endacanthostyles (long acanthostyles originating in endozone) absent (0); present (1).
46. Mural styles absent (0); present in exozone, stereom, or vesicle roofs (1).
47. Styles absent or noninflecting (0); inflecting apertures (1); producing petaloid apertures (2).
48. Paurostyles (mural styles with transverse laminae regularly crossing the core, and with very thin sheaths) absent (0); present (1).
49. Stenostyles absent (0); sparse (1); abundant (2); organized into stellate clusters (3).
50. Pustules (mural lacunae; granules; small patches of crinkled wall laminae) absent (0); present in exozone (1).
51. Carinal styles absent (0); present in longitudinal rows (1); embedded within a carinal layer (2).

INTRAZOOECIAL STRUCTURES:

52. Complete planar diaphragms absent or rare (0); sparse (1); abundant (2).
53. Curved diaphragms absent (0); sparse (1); abundant (2).
54. Cystiphragms absent (0); sparse (1); abundant (2).
55. Perforate diaphragms (ring septa) absent (0); present (1).
56. Hemiphragms (incomplete diaphragms) absent (0); sparse (1); abundant (2).
57. Inferior (proximal) hemisepta absent or rare (0); sparse (1); abundant (2).
58. Superior (distal) hemisepta absent (0); sparse (1); abundant (2).

Appendix C

Generic Originations, Extinctions, and Diversity for Stenolaemate Suborders of the Paleozoic

Note: Suborders (columns) and stages/series (rows) abbreviated by first three letters; originations, extinctions, diversity are enumerated, in that order, by slashes.

	PAL	CER	FIS	EST	HAL	AMP
Tre	0/0/0	0/0/0	1/0/1	2/0/2	0/0/0	0/0/0
Are	2/0/2	1/0/1	2/1/2	4/1/5	9/0/9	4/0/4
Llv	6/0/8	3/1/4	8/3/11	0/1/4	10/0/19	3/0/7
Lle	1/1/9	4/0/7	3/1/11	0/0/4	8/1/27	3/1/10
Car	5/4/13	1/0/8	3/2/12	2/2/6	9/7/34	8/3/17
Ash	4/4/13	1/5/9	3/5/14	0/3/4	3/9/31	2/3/16
Llo	0/0/9	1/0/5	2/0/3	0/1/1	1/2/23	3/0/16
Wen	0/2/9	0/1/5	2/2/15	0/0/0	2/6/23	2/2/18
Lud	0/1/9	0/1/4	1/3/14	0/0/0	3/7/20	4/3/20
Pri	0/1/6	0/3/3	2/4/13	0/0/0	1/9/14	1/2/18
Loc	0/1/5	0/0/0	1/1/0	0/0/0	0/0/5	0/1/16
Pra	0/1/4		0/0/9	0/0/0	0/0/5	0/1/15
Ems	0/0/3		0/0/9	1/0/1	1/0/6	0/0/14
Eif	1/1/4		10/10/19	0/0/1	0/0/6	1/0/15
Giv	0/2/3		13/13/22	1/2/2	0/2/6	7/8/22
Fra	0/0/1		3/4/12	1/1/1	0/2/4	1/3/15
Fam	0/0/1		3/4/12	1/1/1	0/2/4	1/3/15
Tou	0/0/1		10/5/18	1/1/1	0/0/1	6/5/16
Vis	0/0/1		2/4/15	0/0/0	0/0/1	5/2/16
Ser	0/0/1		2/4/13	0/0/1	2/7/16	
Bas	0/0/1		2/0/11		1/0/1	1/1/10
Mos	0/0/1		0/0/11		0/0/0	1/1/10
Ste	0/0/1		2/3/13			0/0/9
Ass	0/0/1		2/1/12			0/0/9
Sak	0/0/1		3/4/14			2/1/11
Art	0/0/1		3/5/13			1/0/11
Roa	0/0/1		2/5/10			2/0/13
Wor	0/0/1		0/0/5			3/4/16
Cap	0/0/0		2/4/7			0/8/12
Dzh			0/3/3			0/2/4

PTI	TIM	RHA	PHY	FEN
0/0/0		0/0/0	0/0/0	
4/1/4		1/0/1	1/1/1	0/0/0
12/1/15		3/0/4	3/0/3	1/0/1
11/3/25		4/0/8	5/0/8	0/0/1
11/11/33		8/7/16	8/9/16	2/1/3
4/9/26		3/4/12	4/5/11	5/0/7
0/3/17		3/1/11	0/1/6	6/0/13
4/5/18		2/2/12	0/1/5	5/1/18
0/4/13		2/1/12	1/4/5	3/1/20
1/4/10		0/1/11	0/0/1	1/0/20
1/2/7		0/0/10	0/0/1	1/1/21
1/4/6		2/1/12	0/0/1	3/3/23
0/0/2		0/0/11	0/0/1	4/4/24
1/0/3	0/0/0	2/2/13	1/1/2	8/2/28
2/2/5	1/0/1	3/4/14	0/0/1	6/7/32
0/1/3	0/0/1	0/1/10	0/0/1	0/2/25
1/1/3	0/0/1	1/0/10	0/0/1	5/4/28
1/2/3	0/0/1	10/5/20	0/0/1	14/14/38
1/1/2	2/1/3	4/7/19	1/0/2	3/2/27
2/2/3	1/0/3	3/1/15	0/0/2	4/2/29
1/1/2	0/0/3	0/0/14	0/0/2	1/1/28
0/0/1	0/0/3	0/1/14	0/0/2	1/1/28
0/1/1	0/0/3	2/3/15	0/0/2	2/2/29
0/0/0	1/0/4	0/1/12	1/0/3	0/3/27
	0/1/4	1/2/12	2/5/5	4/2/28
	0/0/3	1/2/11	0/0/0	2/0/28
	2/1/5	2/2/11		1/6/29
	3/3/7	1/2/10		1/13/24
	0/4/4	0/3/8		0/8/11
	0/0/0	0/5/5		0/3/3

Note: Llv = Llanvirnian; Lle = Llandeilian; Llo = Llandoverian

Acknowledgments   We thank Roger Cuffey, Daniel Blake and his paleontology seminar, and Douglas Erwin for their reviews of the manuscript.

References

Anstey, R. L. 1986. Bryozoan provinces and patterns of generic evolution and extinction in the Late Ordovician of North America. *Lethaia* 19:33–51.

———. 1987. Astogeny and phylogeny: Evolutionary heterochrony in Paleozoic bryozoans. *Paleobiology* 13:20–43.

———. 1990. Bryozoans. In K. J. McNamara, ed., *Evolutionary Trends,* pp. 232–252. London: Belhaven Press.

Anstey, R. L. and T. G. Perry. 1970. Biometric procedures in taxonomic studies of Paleozoic bryozoans. *Journal of Paleontology* 44:383–398.

———. 1973. Eden Shale bryozoans: A numerical study (Ordovician, Ohio Valley). *Publications of the Museum, Michigan State University, Paleontological Series* 1:1–80.

Anstey, R. L. and S. F. Rabbio. 1989. Regional bryozoan biostratigraphy and taphonomy of the Edenian stratotype (Kope Formation, Cincinnati Area): Graphic correlation and gradient analysis. *Palaios* 4:574–584.

Anstey, R. L., S. F. Rabbio, and M. E. Tuckey. 1987. Bryozoan bathymetric gradients within a Late Ordovician epeiric sea. *Paleoceanography* 2:165–176.

Astrova, G. G. 1978. Istoriya razvitiya, sistema i filogeniya mshanok, otriad Trepostomata. *Akademiya Nauk S.S.S.R., Trudy Paleontologicheskogo Instituta* 169:1–240.

Blake, D. B. 1983. The Order Cryptostomata. In R. S. Boardman et al., eds., *Treatise on Invertebrate Paleontology, Part G, Bryozoa Revised, Volume 1,* pp. 440–452. Lawrence, Kansas: Geological Society of America and University of Kansas Press.

Boardman, R. S. 1984. Origin of the post-Triassic Stenolaemata (Bryozoa): A taxonomic oversight. *Journal of Paleontology* 58:19–39.

Boardman, R. S., A. H. Cheetham, D. B. Blake, J. Utgaard, O. L. Karklins, P. L. Cook, P. A. Sandberg, G. Lutaud, and T. S. Wood. 1983. *Treatise on Invertebrate Paleontology, Part G, Bryozoa Revised, Volume 1.* Lawrence, Kansas: Geological Society of America and University of Kansas Press.

Brown, G. D. and E. J. Daly. 1985a. Analysis of evolution of structural characters of *Parvohallopora* Singh from the Dillsboro Formation, Cincinnatian Series, Indiana. In C. Nielsen and G. Larwood, eds., *Bryozoa: Ordovician to Recent,* pp. 51–58. Fredensborg, Denmark: Olsen and Olsen.

———. 1985b. Trepostome Bryozoa from the Dillsboro Formation (Cincinnatian Series) of southeastern Indiana. *Indiana Geological Survey Special Report* 33:1–95.

Cuffey, R. J. 1967. Bryozoan *Tabulipora carbonaria* in Wreford Megacyclothem (Lower Permian) of Kansas. *University of Kansas Paleontological Contributions, Bryozoa, Article* 1:1–96.

————. 1973. An improved classification, based upon numerical-taxonomic analyses, for the higher taxa of entoproct and ectoproct bryozoans. In G. P. Larwood, ed., *Living and Fossil Bryozoa*, pp. 549–564. London: Academic Press.

Cuffey, J. and D. B. Blake. 1991. Cladistic analysis of the Phylum Bryozoa. In F. P. Bigey, ed., *Bryozoaires Actuels et Fossiles: Bryozoa Living and Fossil. Bulletin de la Societe des Sciences Naturelles de l'Ouest de la France* Hors Serie 1:97–108.

Cutler, J. F. 1968. Morphology, taxonomy, and evolution of the bryozoan *Constellaria* from the Cincinnati Arch. Ph.D. dissertation, Columbia University.

Farmer, J. D. and A. J. Rowell. 1973. Variation in the bryozoan *Fistulipora decora* (Moore and Dudley) from the Beil Limestone of Kansas. In R. S. Boardman, A. S. Cheetham, and W. A. Oliver, eds., *Animal Colonies*, pp. 377–394. Stroudsburg, Penn.: Dowden, Hutchinson, and Ross.

Foote, M. and S. J. Gould. 1992. Cambrian and Recent morphological disparity. *Science* 258:1816.

Horowitz, A. S. 1970. Nomenclatorial diversity within *Fistulipora* McCoy and allied genera. *Journal of Paleontology* 44:776–778.

Horowitz, A. S. and J. F. Pachut. 1993. Specific, generic, and familial diversity of Devonian bryozoans. *Journal of Paleontology* 67:42–52.

I Iu, Z.-X. and N. Spjeldnaes. 1991. Early Ordovician bryozoans from China. In F. P. Bigey, ed., *Bryozoaires Actuels et Fossiles: Bryozoa Living and Fossil. Bulletin de la Societe des Sciences Naturelles de l'Ouest de la France, Memoire* Hors Serie 1:179–185.

Key, M. M. 1987. Partitioning of morphologic variation across stability gradients in Upper Ordovician trepostomes. In J. P. Ross, ed., *Bryozoa: Present and Past*, pp. 142–152. Bellingham, Washington: Western Washington University.

Lande, R. 1985. Expected time for random genetic drift of a population between stable phenotypic states. *Proceedings of the National Academy of Sciences of the U.S.A.* 82:7641.

Maddison, W. P. and D. R. Maddison. 1992. *MacClade: Analysis of Phylogeny and Character Evolution, Version 3.0*. Sunderland, Mass.: Sinauer.

Pachut, J. F. 1982. Morphologic variation within and among genotypes in two Devonian bryozoan species: An independent indicator of paleostability? *Journal of Paleontology* 56:703–716.

————. 1987. Population genetics of four species of Ordovician bryozoans: Stereology and jackknifed analysis of variance. *Journal of Paleontology* 61:927–941.

————. 1989. Heritability and intraspecific heterochrony in Ordovician bryozoans from environments differing in diversity. *Journal of Paleontology* 63:182–194.

Pachut, J. F. and R. J. Cuffey. 1991. Clinal variation, intraspecific heterochrony, and microevolution in the Permian bryozoan *Tabulipora carbonaria*. *Lethaia* 24:165–185.

Pachut, J. F., R. J. Cuffey, and R. L. Anstey. 1991. The concepts of astogeny and ontogeny in stenolaemate bryozoans, and their illustration in colonies of *Ta-*

bulipora carbonaria from the Lower Permian of Kansas. *Journal of Paleontology* 65:213–233.

Palmer, A. R. 1986. Inferring relative levels of genetic variability in fossils: The link between heterozygosity and fluctuating asymmetry. *Paleobiology* 12:1–5.

Perry, T. G. and D. B. Hattin. 1958. Astogenetic study of fistuliporoid bryozoans. *Journal of Paleontology* 32:1039–1050.

Prezbindowski, D. R. and R. L. Anstey. 1978. A Fourier-numerical study of a bryozoan fauna from the Threeforks Formation (Late Devonian) of Montana. *Journal of Paleontology* 52:353–369.

Ross, J. P. 1967. Evolution of ectoproct genus *Prasopora* in Trentonian time (Middle Ordovician) in northern and central United States. *Journal of Paleontology* 41:403–416.

Schafer, P. and E. Fois-Erickson. 1986. Triassic Bryozoa and the evolutionary crisis of Paleozoic Stenolaemata. In O. Walliser, ed., *Global Bio-Events,* pp. 251–255. Berlin: Springer Verlag.

Simpson, G. G. 1944. *Tempo and Mode in Evolution.* New York: Columbia University Press.

Taylor, P. D. and G. B. Curry. 1985. The earliest known fenestrate bryozoan, with a short review of Lower Ordovician Bryozoa. *Palaeontology* 28:147–158.

Tuckey, M. E. 1988. Global biogeography, biostratigraphy and evolutionary patterns of Ordovician and Silurian Bryozoa. Ph.D. dissertation, Michigan State University.

———. 1990a. Biogeography of Ordovician bryozoans. *Palaeogeography, Palaeoclimatology, Palaeoecology* 77:91–126.

———. 1990b. Distribution and extinctions of Silurian Bryozoa. *Geological Society of London, Memoir* 12:197–206.

Tuckey, M. E. and R. L. Anstey. 1989. Gradient analysis: A quantitative technique for biostratigraphic correlation. *Palaios* 4:475–479.

———. 1992. Late Ordovician extinctions of bryozoans. *Lethaia* 25:111–117.

Utgaard, J. and T. G. Perry. 1964. Trepostomatous bryozoan fauna of the upper part of the Whitewater Formation (Cincinnatian) of eastern Indiana and western Ohio. *Indiana Geological Survey Bulletin* 33:1–111.

Valentine, J. W. and D. H. Erwin. 1987. Interpreting great developmental experiments: The fossil record. In R. A. Raff and E. C. Raff, eds., *Development as an Evolutionary Process,* pp. 71–107. New York: Liss.

9

Coordinated Stasis and Evolutionary Ecology of Silurian to Middle Devonian Faunas in the Appalachian Basin

Carlton E. Brett and Gordon C. Baird

Speciation clearly plays a key role in punctuational evolution (Eldredge and Gould 1972; Stanley 1979; various papers in this volume), but an equally significant part of the model of punctuated equilibria is the concept of stasis (Gould and Eldredge 1977). Most previous studies have considered stasis and punctuational speciation in the context of single species such as *Phacops rana* or the terrestrial gastropod *Cerion* (e.g., Eldredge and Gould 1972). However, there is increasing evidence that the times of evolutionary stasis among numerous unrelated lineages within communities may be coordinated to a large degree; that is, many different lineages and entire communities of organisms display little or no change for most of their evolutionary history.

For example, Williamson (1981) reported nearly synchronous episodes of morphological change, following much longer intervals of stability, in several lineages of molluscs from Lake Turkana, Africa. Vrba (1985) has documented a comparable pattern of stability and pulses of change in African bovids and hominids; she attributes the abrupt changes to major climatic fluctuations that broke down stable plant and mammal community structures and facilitated rapid evolutionary change and immigration. Change at the level of species as well as entire interacting groups of organisms (i.e., biofacies or communities) appears to occur more or less synchronously. In this paper, we present evidence that a lack of significant evolutionary change is ubiquitous and characteristic of both species and entire biofacies, at least in the middle Paleozoic of the Appalachian Basin.

Ecological-Evolutionary Units and Subunits

As long ago as the 1840s, Alcide d'Orbigny (1849, pp. 299, 382–841; 1850, pp. 190, 197, 394, 427), the "father of biostratigraphy," recognized a pattern of stratal packages which he termed "étages." With considerable insight, he argued that within these relatively large intervals of time, most genera and component species were unchanging. Although this research was done in a pre-evolutionary context, other authors have more recently expanded upon this theme (Boucot 1975, 1983, 1990a, 1990b; Sheehan 1985). In particular, Boucot has proposed that the Phanerozoic can be subdivided into about twelve "ecological-evolutionary units." In each of these units, time series of evolving communities—the so-called "community groups" (Boucot 1982)—display little change, at least at family levels. These ecological-evolutionary units are punctuated by major extinction events and evolutionary radiations that greatly disrupt community structure. Boucot (1990a, 1990b) also argues that, following extinction events, level bottom communities rediversify rapidly by processes of quantum evolution. During the long interludes of stability of an ecological-evolutionary unit, little substantial change takes place in diversity or relative abundance of most lineages within community groups. This pattern appears to be similar to the phenomenon referred to as *biomeres* by Palmer (1965, 1979, 1984) for Cambrian to Early Ordovician trilobite faunas.

Boucot suggests that, despite this pattern of overall stasis, less common lineages may nevertheless undergo slight phyletic evolution that produces species-level changes within ecological-evolutionary units. For the very rare members of community groups, even greater rates of evolutionary change are inferred, which may lead to the production of subgenera or genera as well as new species within the span of one ecological-evolutionary unit. To some degree this pattern may represent a sampling artifact (Koch and Morgan 1988); very intensive collecting, for example, in the Devonian in New York has greatly extended the ranges of several rare species. However, Boucot's notion is now being tested more rigorously with morphometric analysis of rare and common species in the Devonian of the Appalachian Basin.

Boucot's (1990a, 1990b) and Sheehan's (1985, 1992) concept of an ecological-evolutionary unit encompasses a relatively long time interval. For example, ecological-evolutionary unit V of Sheehan (1985) spans Early Silurian to Middle-Late Devonian time—approximately 70 million years—and is bounded by major mass extinctions of the Ashgillian and Frasnian-Famennian boundary. Boucot (1988) divides this same interval into two ecological-evolutionary units: one in the Early Silurian and one encompassing the remainder of the time.

Our paper focuses on Sheehan's ecological-evolutionary unit V. Detailed study of Appalachian Basin Silurian-Devonian biotas supports and builds upon the concepts of Boucot and Sheehan. In particular, however, it is now apparent that ecological-evolutionary unit V is subdivisible into smaller, more fundamental units of time in which species- and community-level characters change very little.

The postulates of Boucot and Sheehan for community group stability and eventual collapse and reorganization apply with equal, if not greater, force to shorter duration intervals, usually 3–7 million year subdivisions within the larger ecological-evolutionary units. During these intervals, even rare species undergo, at most, only species-level changes. Most taxa persist with little obvious morphological change at all, and very few lineages become extinct. Consequently, we term these shorter intervals of community stability *ecological-evolutionary subunits*, or E-E subunits (Brett and Baird 1992; also see Boucot 1990c). We recognize that these intervals are analogous to "biomeres," as presently defined by Palmer (1979, 1984), but we opt for the more general term (E-E subunits) because of the lingering, if incorrect, implication that biomeres are nonevolutionary and perhaps diachronous. Furthermore, it appears that the boundaries of several of the finer subdivisions now identified in the Appalachian Basin may coincide with minor extinction events such as those recognized within the Siluro-Devonian interval by Boucot (1990b).

During a typical E-E subunit, at least 65% and typically more than 80% of morphospecies appear to persist with little or no morphological change. In these intervals, biofacies retain their identity, but major collapse and restructuring events occur at their boundaries. We refer to the phenomenon of shared stability during E-E subunits as "coordinated stasis." The same term could be used for the blocks of stability in Vrba's (1985) "stability-pulse hypothesis."

Ecological-Evolutionary Subunits in the Silurian to Middle Devonian Interval, Appalachian Basin

The Silurian to Middle Devonian interval in the northern Appalachian Basin is among the best studied stratigraphically and paleontologically in North America. Traditionally, this interval has been subdivided into a set of local series and stages, which have largely been replaced in recent times by European stage terminology (figures 9.1 and 9.2). The present contribution, based as it is on detailed biostratigraphic work of numerous earlier researchers and our own research, affirms the detailed bioevent zonation of the Appalachian Basin strata. More important, it serves to illustrate the concept of ecological-evolutionary subunits.

During the past two decades we have compiled data on the occurrence of fossil species and biofacies within an increasingly refined sequence and event stratigraphic framework in the Silurian to Middle Devonian rocks of New York State, Pennsylvania, and Ontario (Brett 1986; Landing and Brett 1991; Brett, Miller, and Baird 1990). This interval is bracketed, in part, by major extinctions in the Ashgillian (Late Ordovician) and in the Frasnian (Late Devonian). However, there are evidently a number of minor crises involving at least local extinctions within this block of time. The rocks represent a time span of approximately 65 or 70 million years (Harland et al. 1982), which has been subdivided into 12 to 14 local series or stages. Some of these correspond approximately to the equal number of ecological-evolutionary subunits defined here (figures 9.1 and 9.2).

Faunas typical of each stage or E-E subunit do not show gradual progressive change to the next E-E subunit, but rather abrupt breaks with the overlying and underlying faunas. While minor disconformities occur at some of these boundaries, they appear to represent a small fraction of the time recorded by strata within the intervals of stability. Furthermore, the best documented E-E subunits display nearly conformable boundaries, as we shall discuss. Radiometric dating in the Silurian to Middle Devonian interval is very poor, and therefore only tentative assignments of duration for various subunits are possible. In this paper we have used the values of McKerrow, Lambert, and Cocks (1985), with some modification to accommodate a few more accurate dates from the Appalachian Basin (e.g., Roden, Parrish, and Miller 1990).

The present compilation is based on a survey of major faunal lists of a relatively small number of paleontologists who have worked extensively

SYSTEM	LOWER SILURIAN					UPPER SILURIAN		
SERIES (NA)	ALEXANDRIAN	NIAGARAN				CAYUGAN		
SERIES (E)	LLANDOVERY			WENLOCK	LUDLOW		PRIDOLI	
STAGES	RHUDDANIAN	AERONIAN	TELYCH.	SHEIN. / HOMER.	GOR. / LUDF.		—	
UNITS (NY)	MEDINA	L. CLINTON		U. CLINTON	LOCKPORT	SALINA	BERTIE RONDOUT	
LITHOLOGIES	red to gray sh., ss., minor ls.	green, purple sh., ls., hem.		gray sh., ss., ls., dol.	dol, ls., dol. shale	red sh., gray dol. sh., dolostone, evaporites		
UNITS CENTRAL APP.	TUSCARORA	ROSE HILL		MIFFLINTOWN	McKENZIE	BLOOMSBURG TONOLOWAY	KEYSER	
LITHOLOGIES	ss. and congl.	green-purple sh., minor ss., hem.		gray ss., shale, minor ls.	dark gray shale, ls.	red, green, gray sh., ls., dol.	ls. and gray sh.	
EE SUBUNITS (duration in millions of yrs)	1. MEDINA (5 Ma.)	2. L. CLINTON (4 Ma.)		3. U. CLINTON - LOCKPORT (7 - 8 Ma.)		4. SALINA (3-4 Ma.)	5. KEYSER (2 Ma.)	
TYPICAL COMMUNITY TYPES	BA- 2: pteriniid 3: *Cryptothyrella* 4: (unnamed diverse Manitoulin Ls. communities) —	2: *Eocoelia* 3: *Pentamerus* 4: *Stricklandia* —		2: *Howellella* - pteriniid 3: *Whitfieldella* - lg. *Leptaena* 4A: *Dicoelosia* - *Skenidioides* 4B: *Striispirifer* - *Amphistrophia* 5: monograptid - *Protochonetes*		2: *Howellella* 3: *Eccentricosta* (no detailed subdivisions)	2: *Howellella* 3: Gypidulinid 4: Keyser Ls. bryozoan - cystoid association	

Figure 9.1 Silurian of the Appalachian Basin: temporal relationships of lithostratigraphic units, ecological-evolutionary (E-E) sub-units, and the typical community types of high diversity calcareous facies. Durations are approximate, based on dates of McKerrow, Lambert, and Cocks (1985). Abbreviations for stages: Rhuddan = Rhuddanian; Aeron = Aeronian; Telychi = Telychian; Shein = Sheinwoodian; Homer = Homerian; Gor = Gorstian; Ldf = Ludfordian; NA = North American series; E = European series/stages. Lithologies: dol = dolostone; hem = hematite; ls = limestone; sh = shale; ss = sandstone; BA = benthic assemblage. For descriptions of named communities see Boucot 1975; Brett, in press (for Wenlockian communities).

SYSTEM	LOWER DEVONIAN			MIDDLE DEVONIAN				U.D.
SERIES (NA)	ULSTERIAN			ERIAN				SEN.
STAGES (NA)	HELDERBERGIAN	DEERPARK	SAWKILL	SOUTHWOOD	CAZ.	TIOUG.	TAG.	F.L.
STAGES (E)	LOCHKOVIAN	PRAGIAN	EMSIAN	EIFELIAN	GIVETIAN			FRAS.
UNITS (NY)	HELDERBERG	ORISKANY	ESOPUS SCHOHARIE	ONONDAGA	HAMILTON	TULLY		GEN.
LITHOLOGIES	micritic, cherty to argillaceous ls., minor shales	qtz. arenite, sandy ls., dk. mudstone	argillaceous ls., calc. shale	cherty ls., minor dk. shale	black-med. gray sh., siltstone, ss., minor ls.	ls., calc. shale		black, gray sh., siltstone
UNITS CENTRAL APP.	HELDERBERG	ORISKANY-RIDGELY	lower NEEDMORE	SELINSGROVE	MARCELLUS-MAHANTANGO	TULLY		BURKET-HARRELL
LITHOLOGIES	micritic, cherty calcarenite, argillaceous ls.	quartz arenite, sandy ls.	med. gray to black calc. sh., minor ls.	micritic ls., calcareous shales	black shale, gray siltstone, ss.	calc. shale		black to greenish gray sh.
E-E SUBUNITS (duration millions of yrs)	6. HELDERBERG (6 Ma.)	7. ORISKANY (2-3 Ma.)	8. SCHOHARIE (5 Ma.)	9. ONONDAGA (5-6 Ma.) [9A SH (1)]	10. HAMILTON (6-7 Ma.)			LL — not subdivided
TYPICAL COMMUNITY TYPES	BA- 2: Tentaculites 3: Gypidulinid (Cyrtina) 4A: Megakozlowskiella 4B: Hedeina 5: Coelospira - Pacificocoelia	2: Hipparionyx 3: Hipparionyx 4A: Plicoplasia - 4B: Leptocoelia 5: ------	2: Chonostrophia 3: Aemulophyllum 4: Etymothyris - Amphigenia 5: Pacificocoelia	2: ------ 3: Heliophyllum 4A: Diverse brach* 4B: Arrypa 5: Pacificocoelia	2: ------ 3: Pentamerella - Helio. 4A: Diverse brachiopod* 4B: Athyris - Medio. 5: Ambocoelia - chonet.			not subdivided

Figure 9.2 Early to Middle Devonian of the Appalachian Basin: temporal relationships of lithostratigraphic units, ecological-evolutionary (E-E) subunits, and the typical community types of high diversity calcareous facies. Abbreviations for series and stages: Caz = Cazenovian; Tioug = Tioughniogan; Tag = Taghanic; F.L. = Finger Lakesian; Fras. = Frasnian. Lithologies: same abbreviations as for Silurian. Subfauna 9A is the short-lived "Stoney Hollow fauna," characterized by an unusual mixture of holdover Onondaga species and the brief appearance of unique forms such as *Variatrypa arctica*. For descriptions of named Lower Devonian communities see Boucot 1982; for Onondaga fauna see Koch 1978, and Feldman 1980; for Hamilton fauna see Brett, Miller, and Baird 1990. *Aemulophyllum* and *Heliophyllum halli* coral associations are discussed in Oliver and Sorauf 1981.

on the Silurian-Devonian faunas. We have attempted to use the work of only a few experienced researchers to maximize consistency in species identification. Also, despite obvious problems of subjectivity in species recognition, we suspect that most of the mid twentieth century biostratigraphic studies which have been used were biased toward splitting species based on minor differences, rather than lumping them. Hence, these data, if anything, are biased against the detection of patterns of stasis.

Detailed work on faunal assemblages in the Silurian is based on Bolton (1957) for Ontario Silurian rocks; Gillette (1947) for the Clinton Group in New York State; Zenger (1965, 1971) for the Lockport Group in New York State; Swartz (1913) for the Silurian and Early Devonian of Pennsylvania, Maryland, and West Virginia; Grabau (1906) and Goldring (1935, 1943) for the Helderberg, Oriskany, and Schoharie formations in eastern New York State; Buehler and Tesmer (1963) for compilations on the Onondaga and Hamilton groups in western New York State; and Heckel (1973) for the Tully Limestone. However, the general observations of coordinated stasis come from our personal observations made during more than two decades of field work in the Silurian and Devonian faunas of New York, Pennsylvania, and Ontario. Comparing faunas of different ages, we have attempted to consider the most comparable ranges of environments from shallow to mid-shelf (benthic assemblages 2 through 4 in Boucot 1975) and mixed siliciclastic and carbonate facies.

Major changes occur in a suite of common species within particular successive faunas. These abrupt jumps punctuated extensive stage-level intervals of recurring biofacies (figures 9.1 and 9.2). For example, the Lower Devonian Helderberg fauna (tables 9.1 and 9.2) displays persistent recurring associations of diverse brachiopods, such as the *Dicoelosia-Hedeina* community (Boucot 1982). In contrast, the overlying Glenerie Limestone, with similar facies, displays a distinctive cast of characters. Fewer than 20% of Helderberg species range upward into the Oriskany, and these are mainly species that are uncommon to rare in both faunas; that is, they are not among the dominant species. To document the change, we computed the percentage of all species from a given fauna that are carried over to the next. In the case of the Helderberg fauna, only 7 of 130 (5%) are holdovers from the underlying Keyser fauna, and only 25 (19%) persist into the overlying Oriskany Sandstone and coeval Glenerie Limestone.

Table 9.1
Attributes of the Lochkovian Helderberg Fauna
(E-E Subunit 6) of the Devonian

Taxa	Early Lochkovian (Kalkberg and New Scotland Formations)		Late Lochkovian (Alsen and Pt. Ewen Formations)	
	T : C	% C	T : H	% H
Porifera	1 : 1	100	1 : 1	100
Corals	4 : 4	100	7 : 4	57
Bryozoans	16 : 14	88	14 : 14	100
Brachiopods	45 : 31	69	45 : 31	69
Bivalves	3 : 3	33	1 : 1	100
Trilobites	5 : 3	60	4 : 3	75
TOTAL*	74 : 54	73	72 : 54	75

NOTE: T = total number of species, tabulated from Grabau (1906) and Goldring (1935, 1943); C = species that carry over from lower to upper cycles; H = species in upper cycle that are holdovers from below. Facies are all siliceous to cherty nodular wackestone and packstone, representing outer shelf, Benthic Assemblage 4 to 5, conditions in the lower and upper major cycles of the Helderberg Group.

*If one combines the Early and Late Lochkovian formations, the total Helderberg fauna comprises 130 species, with an overall persistence of 70% (91/130). The holdover species from the older Keyser fauna account for 5.5% (7/130); 26.6% (25/94) of the upper Helderberg species carryover to the Oriskany.

Table 9.2
Persistence and Extinction Values of Four Well-Characterized Faunas
of the Silurian to Middle Devonian Interval

Fauna	E-E Subunit	Time Span	Persistence	Extinction
Hamilton	10	5–6 My	80%	5%
Onondaga	9	6–7 My	78%	< 10%
Helderberg	6	7–8 My	70%	< 10%
U. Clinton/Lockport	3	7–8 My	66%	32%

NOTE: Persistence denotes the proportion of species that range (within appropriate facies) from the lowest to the highest parts of the fauna (or E-E subunit) in question. Extinction denotes the proportion of species that become extinct prior to the end of the fauna (or E-E subunit).

We also tested for persistence of species within an ecological-evolutionary unit. In the case of the Helderberg faunas, we compared the species content of very similar facies in two major (third order) cycles. The argillaceous deeper shelf limestones of the Kalkberg and New Scotland formations of the lower cycle are very similar to those of the Alsen and Port Ewen beds in the upper (see Laporte 1969). The biofacies also are superficially quite similar, and, indeed, about 54 out of 74 species of invertebrates (or 73%) persist with little or no morphological change from the lower to the higher cycle. Of the remaining 30 taxa, more than half of the lineages are represented by congeneric species with only fairly slight differences in morphology that may have evolved from the older forms via anagenesis. Others, including two genera of inarticulates, are represented by closely related forms in higher post-Helderberg beds or simply may not have been discovered in the higher cycle. Only five new genera appear in the upper cycle that are not present in the lower. Hence, there is very little deletion or addition of lineages within the cycles of the Helderberg.

Overall, the persistence of species within ecological-evolutionary subunits (table 9.2) ranges from a low of 66% to more than 80%. The proportion of species lineages that becomes extinct locally *within* an ecological-evolutionary subunit is quite low (table 9.2), although this number is difficult to assess accurately without very careful search for rare species throughout a stratigraphic interval. The extinction percentage is thus probably overestimated, rather than underestimated, in poorly studied faunas, as many apparently "extinct" rare species simply may not have been discovered yet in upper parts of their ranges.

In the best-known cases of the Onondaga and Hamilton faunas, for example, fewer than 10% of species appear to have become extinct within the subunit. Other faunas are less well studied but display on the order of 10% extinction, or less. We are not including pseudoextinctions, but only termination of lineages; however, pseudoextinction also appears to be quite low in most cases.

The degree of species or faunal persistence between E-E subunits apparently varies somewhat, both among facies and among different major taxa. For example, in comparing Early Silurian Medina and lower Clinton faunas (E-E subunits 1 and 2), a substantially higher carryover of species (44%) is observed in nearshore siliciclastics than in shallow shelf carbonates (36%; table 9.3). Similarly, *within* the upper Clinton (latest Llandovery to Wenlock age; E-E subunit 3), shelly mudstone facies of the outer shelf display a larger persistence of species (78%) than do biostromal shelf carbonates (55%). The latter value, however, is strongly influenced by a large number of silicified coral taxa in the Fossil Hill Formation in the middle portion of the Clinton Group that simply may not have been recognized in the dolomitized reefal facies of the younger Lockport Group (table 9.4).

Table 9.3

Faunal Carryover from Early Silurian Medina (E-E Subunit 1) to Lower Clinton (E-E Subunit 2) in Two Distinct Facies

Facies	E-E Subunit 1	E-E Subunit 2	Carryover
Nearshore Siliciclastic	Cabot Head Shale	Sodus/Sauquoit Shale	
(BA-2)	N = 45	C = 20	44%
Carbonate Shelf	Manitoulin Ls.	Reynales/ Wolcott Ls.	
(BA-3)	N = 94	C = 34	36%
COMBINED	N = 139	C = 54	39%

NOTE: N = total number of species in older fauna; C = number of species that carry over to the younger fauna. BA-2 and BA-3 refer to standard depth-related benthic assemblages as defined by Boucot (1975); BA-2 assemblages represent shallow inner-shelf settings, BA-3 are offshore, near normal wave base.

Table 9.4

Species Carryover from Early to Late Phase of the Upper Clinton/Lockport Fauna (E-E Subunit 3) of the Middle Silurian

Facies Benthic Assemblage	Llandovery C6 (early phase) *total species*	Mid–Late Wenlock (late phase) *species carryover*	Carryover
Carbonate, biostromal facies (BA-3)			
	Fossil Hill Ls.	*Gasport Ls.*	
Corals	50	21	42%
Other taxa	26	21	81%
SUBTOTAL	76	42	55%
Shelly mudstone facies (BA-3, BA-4)			
	Willowvale Shale	*Rochester Shale*	
All taxa	76	59	78%
TOTAL*	152	101	66%

*The shift from the Llandovery C6 to the Mid–Late Wenlock was marked by the extinction of about 32% of the species (48 of the original 152). Most of these were corals. Of the 152 species in the Llandovery, 30 (or about 20%) of these were holdovers from the previous subunit (2), the Lower Clinton.

In any case, for overall faunal comparisons (table 9.5) we have used integrated samples of varied biofacies. Table 9.5 summarizes the attributes of the ten major ecological-evolutionary subunits that we now recognize within the Silurian to Middle Devonian interval in the Appalachian Basin. A series of four or five short duration faunas appear to have existed in the Frasnian Stage (McGhee 1988, 1990), but these have not been studied in detail and so are not included in table 9.6. The ecological-evolutionary subunits will be described in greater detail in another publication (Brett, in preparation).

Percentages in tables 9.5 and 9.6 display the proportion of species for a given fauna that represent holdovers from a previous fauna and the proportion that carries over to the next. The total numbers of species have been compiled for mixed siliciclastic and carbonate facies, primarily in New York State, from our field observations and from a literature search. They represent only a preliminary compilation of macro-fossil

Table 9.5

Appalachian Basin Silurian and Devonian Faunas, Showing Holdover and Carryover Indices.

E-E Subunit	Fauna	Age	Duration	Carryover	Holdover
10	Hamilton-Tully	Givetian	6–7	30/335 (9%)	32/335 (10%)
9	Onondaga*	Eifelian	5	32/200 (16%)	37/200 (18%)
8	Schoharie	Emsian	5	37/125 (30%)	10/125 (8%)
7	Oriskany	Pragian	2	10/94 (11%)	25/94 (26%)
6	Helderberg	Gedinnian	6	25/130 (19%)	7/130 (5%)
5	Keyser-Bertie	Pridolian	2	14/54 (26%)	7/54 (13%)
4	Salina	L. Ludlovian	3–4	7/48 (15%)	14/48 (16%)
3	U. Clinton–Lockport	L. Llandovery–Wenlock	7–8	30/149 (20%)	7/146 (5%)
2	L. Clinton	Mid-Llandovery	4	48/87 (55%)	30/87 (34%)
1	Medina	E. Llandovery	5	?	48/139 (34%)

NOTE: Holdover index is expressed as the ratio of the number of species derived from the previous fauna to the total number of species in the subunit being considered; carryover index is the proportion of species from the considered fauna that continue into the next younger subunit. Durations of the subunits (in millions of years) are estimates based mainly on the time scale in McKerrow, Lambert, and Cocks 1985.

*In this tabulation, the Onondaga fauna includes a distinctive subunit, the Stony Hollow subfauna, which characterizes the lowest portion of the Hamilton Group.

Table 9.6

Within-Facies Comparison of Onondaga
(E-E Subunit 9; Eifelian) and Hamilton (E-E Subunit 10; Givetian)
Faunas of the Middle Devonian

Taxa	CARRYOVER from Eifelian to Givetian	HOLDOVER in Givetian from Eifelian
Corals	9/46 = 19.6%	9/43 = 20.9%
Brachiopods	9/46 = 19.6%	9/60 = 15.0%
All taxa	32/200 = 16.0%	32/335 = 9.5%
Within-fauna persistence*	E-E Subunit 9	E-E Subunit 10
	156/200 = 78.0%	268/335 = 80.0%

NOTE: Data (in numbers of species) are drawn from similar facies in the two E-E subunits: slightly argillaceous, coral-rich wackestone to packstone carbonates. Calculations are based on compilations for the Moorehouse Member of the Onondaga Formation (E-E subunit 9) and for the Centerfield and Tichenor members of the Ludlowville and Moscow formations of the Hamilton Group (E-E subunit 10).
*Denotes the percentage of all species that persist throughout each of the two faunas—from the lowest to the highest occurrence of the appropriate biofacies.

species, and the absolute numbers are subject to minor change with further detailed study. However, we are confident that proportions listed here are approximately correct.

It is evident in all cases that only a relatively small percentage (average 19%) of species carry over from one fauna to the next in relatively conformable successions. In some cases, minor disconformities may occur at the boundaries of these intervals, but in the best-studied examples—that is, the middle Silurian Clinton fauna (E-E subunit 3) and the Middle Devonian Onondaga and Hamilton faunas (E-E subunits 9 and 10)—major faunal changes can be demonstrated to take place within essentially conformable successions. These faunal changes are also abrupt, occurring within the duration of a single, small-scale sedimentary cycle that probably represents no more than a few hundred thousand years.

A Detailed Example: Middle Devonian Hamilton Fauna

The Middle Devonian (Givetian) contains the Hamilton fauna (E-E subunit 10). This subunit is the best studied of the blocks of coordinated stasis known to exist within the Appalachian Basin faunal succession. The duration of the Givetian interval is debatable. Its upper boundary with the Frasnian is rather consistently placed at 374–377 Ma (Harland et al. 1982, 1990; McKerrow, Lambert, and Cocks 1985; Gale 1985), but the base of the Givetian has been considered as young as 381 Ma (Harland et al. 1990). However, a radiometric date of 384 Ma obtained for the lower Hamilton (Marcellus shale) in Pennsylvania (Bofinger and Compston 1967) and a newly published, very precise date of 390 ± 0.5 Ma for a K-bentonite high in the Onondaga Limestone (middle Eifelian; Roden, Parrish, and Miller 1990) argue for a considerably older base for the Givetian and, thus, a duration of up to nine million years for the stage.

In this section we consider the Givetian stage and its attendant Hamilton fauna as a case study of ecological-evolutionary stability. This well-known suite of marine invertebrate fossils occurs in dark gray to black mudrocks, sandstones, and thin carbonates in New York and Pennsylvania. The Hamilton fauna is a part of the Eastern Americas realm, which existed in subtropical to warm temperate shallow environments in eastern Laurentia; it also contains elements derived from the Old World Rhenish-Bohemian province (figure 9.3 and 9.4). Overall, the Hamilton fauna includes more than 330 species of corals, bryozoans, brachiopods, molluscs, trilobites, echinoderms, and other invertebrates which occur in a series of about twenty recurring biofacies or "communities" (Brett, Miller, and Baird 1990).

The Hamilton fauna abruptly replaces a subunit of the older Onondaga fauna near the Eifelian-Givetian stage boundary (Boucot 1990b; Brett, Miller, and Baird 1990, figure 1). The fauna of the underlying Onondaga limestone (E-E subunit 9) comprises more than 200 species, including long-ranging corals, brachiopods, crinoids, and trilobites—most of which disappear abruptly near the Eifelian-Givetian stage boundary, low in the Hamilton Group. Only about 16% of the genera carry through (table 9.6). The overlying Marcellus Formation in west-central New York consists mainly of black shales, a facies inappropriate for comparison with the underlying carbonates. However, in more proximal areas, some Onondaga faunal elements persist into the first major shallowing cycle of the Marcellus (Stony Hollow Sandstone–Cherry Valley Limestone; Griffing and Ver Straeten 1991). These holdovers are mixed with an unusual suite of species (e.g., *Pentamerella winteri* Johnson and *Varitrypa arctica* Warren) that appear to reflect a short-lived incursion (epibole) of warm water taxa from the Old World, possibly from northwestern areas (Koch 1979). These unusual organisms, along with most Onondaga holdovers,

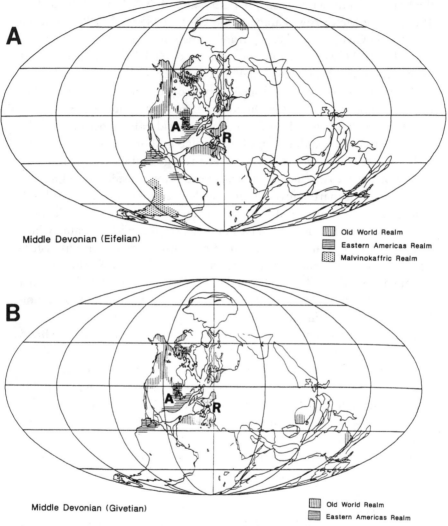

Figure 9.3 Global paleogeography during the Middle Devonian Eifelian and Givetian Stages, showing the relative positions of the Eastern North America and Old World Realm Faunas. A = Appohimchi (Appalachian Basin) Province; R = Rhenish-Bohemian Province. Modified from Blodgett, Rohr, and Boucot 1990; base maps from Scotese 1986.

1 cm

Figure 9.4 Common coral and brachiopod species from the lowest *(left)* and highest *(right)* coral-rich associations sampled in the Middle Devonian Hamilton Group. *Top to bottom: Heterophrentis* cf. *H. simplex; Athyris* cf. *A. cora; Mediospirifer audaculus; Pseudoatrypa* cf. *P. devoniana;* left suite of specimens is from the Halihan Hill Coral bed of the basal Otsego Member, Marcellus Formation, Middleburgh, Schoharie County, New York; the right suite is from the Fall Brook Coral bed of the Windom Shale Member, Moscow Formation, Genesee Valley, Livingston County, New York.

do not reappear in the second shoaling cycle of the Hamilton Group, which culminates in a coral-rich bed. Rather, the typical Hamilton biota appears in place of the Stony Hollow or typical Onondaga faunas.

Comparing only the faunas of the slightly argillaceous packstones and wackestones in the Moorehouse Member of the Onondaga Formation with those of very similar facies in the later Centerfield and Tichenor Limestone members in the Hamilton Group (table 9.6), we note that 9 out of 46 Onondaga corals persist (19.6%; these 9 comprise about 21% of the 43 Hamilton taxa). Among brachiopods the result is virtually identical; 19.5% are found to carry over, forming about 15% of the more diverse brachiopod assemblage of the Hamilton carbonate facies. Considering all species, including corals, brachiopods, trilobites, molluscs, echinoderms, and others, there is approximately a 16% carryover from the Onondaga into the Hamilton Group (table 9.5). These species comprise less than 10% of the total Hamilton fauna. Yet, both the Onondaga and Hamilton faunas display considerable stability. About 78% of all species persist in approximately similar facies from the lowest to the highest samples of the Onondaga, and at least 80% of species occur throughout the entire Hamilton-Tully interval.

Of the more than 300 non-Onondaga species in the Hamilton fauna, only a small proportion are derived from preexisting lineages within the Appalachian Basin. Instead, about half of the Hamilton fauna appears to record immigrant taxa. Many of them—including, for example, the familiar orthid brachiopod *Tropidoleptus carinatus*, many of the bivalves, and the well-known phacopid trilobite *Phacops rana*—appear to have arisen during earlier Devonian times in the Old World, Rhenish-Bohemian realm (Boucot 1990a; see also McIntosh 1983 for crinoids). These species appear abruptly in a single, very widespread, fossil-rich, gray mudstone horizon (Halihan Hill bed of Ver Straeten 1993) in the Marcellus Shale of the Hudson Valley area (previously termed the *Meristella*-coral bed [Cooper, in Goldring 1943]). We have recently discovered this assemblage in a 20–30 cm gray mudstone in the otherwise black shale facies of the Oatka Creek Shale Member of the Marcellus Formation in western New York. This bed occurs slightly above the Cherry Valley Limestone, itself near the Eifelian-Givetian stage boundary (Klapper 1981). The Cherry Valley, like the slightly older Stony Hollow, carries an unusual fauna dominated by pelagic cephalopods and styliolinids, plus a few benthic faunal elements that seem to belong in neither typical Hamilton nor Onondaga faunas. However, the underlying Stoney Hollow Sandstone and its lateral equivalents (Chestnut Street bed of Griffing and Ver Straeten 1991), carries a number of last occurrences of Onondaga taxa, such as the brachiopods *Leptaena rhomboidalis* and *Pacificocoelia*. Thus, in conformable successions, the Hamilton fauna occurs in shallow water facies within one small-scale cycle (fifth order, probably a few hundreds of thousands of years duration [Brett and Baird 1985, 1986, 1992]) above the last appearance of some typical Onondaga forms.

Individual lineages within particular biofacies of the Hamilton biotas appear to display very little morphologic change, and that which is observed is neither progressive nor directional. This phenomenon was first noticed by Cleland (1903), who made a detailed (though nonstatistical) survey of populations of more than two hundred species of invertebrate fossils from beds at all levels in the Hamilton Group in the Cayuga Lake area. Cleland makes a perfunctory and obviously wistful statement in the discussion that he was unable to find a single good example of progressive evolutionary change among these Hamilton species, despite careful search.

Morphometric analysis of several Middle Devonian lineages is ongoing (e.g., Lieberman, Brett, and Eldredge, in press). A detailed discussion of the results is beyond the scope of this paper. It should be noted, however, that careful study of several lineages by taxonomic specialists has revealed nearly complete morphological stasis or very minor and nondirectional phyletic change in nearly all cases. To date, studied taxa include species of corals, bryozoans, brachiopods, and trilobites. Most studies, for practical reasons, have dealt with relatively common taxa—that is, those found in abundance (on the order of five or more specimens per square meter of bedding plane) in nearly all samples. Preliminary, unpublished studies of rare species of trilobites, crinoids, and brachiopods nevertheless reveal similar patterns of nearly complete stasis. Species studied range from stenotopic corals to nearly ubiquitous trilobites.

Eldredge's (1972) thorough documentation of evolution in the Middle Devonian trilobite *Phacops rana* (Green) is the classic example of punctuational change. More significant, but less well known, is the fact that this same study provides an excellent example of nearly complete stasis. Eldredge examined more than fifty characters in *P. rana*, but he was able to document a progressive evolutionary change in only one feature: the relatively trivial reduction from 18 to 16 vertical files of lenses in the eyes.

That reduction involved just two punctuational steps over the course of five to six million years. Morphometric studies of two other trilobites, *Dipleura dekayi* (Green) and *Basidechenella rowi* (Green), also indicate morphologic stasis in the Hamilton interval (Lieberman 1993, and personal communication).

Pandolfi and Burke (1989a, 1989b) made a detailed study of ecophenotypic and possible long-term evolutionary change in two Hamilton tabulate corals. Using Fourier analysis, they found some facies-related variation in colony shape, but no significant directional change in samples obtained from low and high in the Hamilton Group. It is notable that *Pleurodictyum dividua* (Hall), a rare and very stenotopic coral, displayed no more evolutionary change than did the common *P. americanum* (Roemer). Pandolfi and Burke (1989b) conclude that both corals display morphological stasis throughout the Hamilton interval. Similarly, intensive study of the rugose coral *Heliophyllum* has led Sorauf and Oliver (1976; also personal communication, 1993) to conclude that all specimens from the Hamilton Group are assignable to the single species *H. halli* (Edwards and Haime). They document minor random variation in septal carinae microarchitecture, but that variation does not appear to be progressive.

Quantitative morphometric studies of Hamilton fossils have also demonstrated effective stasis in several brachiopod lineages. Welch (1991) demonstrated considerable morphological plasticity in *Mucrospirifer mucronatus* (Conrad), but no directional change within Hamilton populations. Similarly, Lieberman, Brett, and Eldredge (in press) used multivariate principal components and canonical discriminant analysis to test for morphological change in eight parameters in the shells of the common brachiopods *Athyris spiriferoides* (Eaton) and *Mediospirifer audaculus* (Conrad). Surprisingly, populations from the lowest and highest samples in the Hamilton Group resembled one another most closely. Minor, seemingly random, variations occurred in intervening samples, but these were not consistently associated with either age or facies. Goldman and Mitchell (1990) concluded that two species of ambocoeliid brachiopod, *Ambocoelia umbonata* (Conrad) and *Crurispina nana* (Grabau), displayed consistent morphologies throughout most of the Hamilton Group, while a third lineage with very sporadic occurrence, *Emanuella*, may have undergone a species-level change from *E. subumbona* (Hall) to *E. praeumbona* (Hall) in the late Givetian. Isaacson and Perry (1977) were unable to find any significant morphological change in the common Hamilton

Group brachiopod *Tropidoleptus carinatus* (Conrad) from its lowest to its highest occurrence, a total span of some 40 million years (including all of the Givetian Stage); detailed morphometric analysis of this species, ongoing, suggests the same pattern (Eldredge, personal communication).

Only a single detailed study, involving trepostome bryozoans (*Leptotrypella*), suggests the evolution of two or more species during deposition of the Hamilton Group—in this case, both speciations occurred in the upper part (Boardman 1960). However, the wide stratigraphic separation of these populations precludes study of the evolutionary mechanisms. The new species may have arisen by steady (or episodic) anagenesis or by more abrupt speciation.

Taken together, these studies indicate that a majority of Hamilton lineages display virtual stasis from oldest to youngest samples. Slight nondirectional change is observed in some cases. Such variation seemingly records very minor evolutionary fluctuation of the sort documented by Sheldon (1987, 1993). However, it clearly does not lead to development of major new grades of morphological development, nor to ecological expansion for any of the taxa involved. Several of the species appear abruptly in the Appalachian Basin near the beginning of the Hamilton fauna or become locally extinct at its end.

A more detailed picture emerges if we focus on changes or the lack of changes within particular biofacies or "communities" throughout the Hamilton subunit in the Appalachian Basin. The Hamilton fauna is divisible into a number of persistent faunal associations or biofacies that were controlled by variations in water depth and sedimentation (Brett, Miller, and Baird [1990] and Brower and Nye [1991] provide further discussion). These biofacies recur cyclically in the five to six million year span of the Givetian, providing an opportunity to examine evolution within habitats.

Both common and relatively rare species from the lowest to highest known coral-rich beds of the Hamilton are very similar (table 9.7). The coral beds were chosen for comparison because they contain the highest diversity faunas in the Hamilton's mixed carbonate-siliciclastic facies and also because they contain a large number of stenotopic species that occur in no other facies. Not only the individual species but also the biofacies appear to persist.

Table 9.7 also compares the dominant and moderately common taxa in the two coral beds. Both contain about sixty to seventy common fossil

Table 9.7

Comparison of Common Species of the Lowest and Highest Known Coral Beds in the Hamilton Group of New York State.

MERISTELLA-CORAL BED Lower Mt. Marion (= Marcellus) Fm.	FALL BROOK CORAL BED Windom Member; Moscow Fm.

AGE:
Earliest Givetian; *Pol. ensensus* conodont zone; (~384 Ma)

Late Givetian; middle *Pol. varcus* conodont zone; (~378 Ma)

TYPICAL LOCALITIES:
Berne area, Albany Co., N.Y.
Middleburgh, Schoharie Co., N.Y.

Fall Brook; Little Beards Creek;
Geneseo area, Livingston Co., N.Y.

REFERENCE:
Goldring (1935)

Baird and Brett (1983)

DOMINANT CORALS:
Aulocystis sp. (1)
Amplexiphyllum sp. (2)
Cystiphylloides cf. *C. americanum* not present ?
Favosites spp. (small coralla)
Heliophyllum cf. *H. halli*
Heterophrentis cf. *H. simplex* (3)
Stereolasma rectum

Aulocystis sp. (1)
Amplexiphyllum hamiltoniae (2)
Cystiphylloides americanum (3)
Eridophylum subaesptosum
Favosites spp. (small coralla)
Heliophylum halli
Heterophrentis simplex (4)
Stereolasma rectum

DOMINANT BRACHIOPODS:
Athyris cora (2)
"*Camarotoechia*" *prolifica*
Coelospira cf. *C. camilla*†
Cyrtina hamiltoniae
Elita fimbriata
Mediospirifer audaculus (1)
Mucrospirifer mucronatus
Protodouvillina inequistriata (3)
Protoleptostrophia perplana
Pseudoatrypa cf. *P. devoniana* (4)
Rhipidomella cf. *R. vanuxemi*
not present
Spinulocosta spinulocosta

Athyris spiriferoides
"*Camarotoechia*" spp. (2)
not present
Cyrtina hamiltoniae
Elita fimbriata
Mediospirifer audaculus (1)
Mucrospirifer mucronatus
Protodouvillina inequistriata (3)
Protoleptostrophia perplana
Pseudoatrypa cf. *P. devoniana* (4)
Rhipidomella cf. *R. vanuxemi* (5)
Spinatrypa spinosa* (2)
Spinulocosta spinulocosta

OTHER BRACHIOPODS:
Ambocoelia cf. *A. umbonata*
Devonochonetes coronatus
Eumetabolotoechia cf. *E. multicostum*
Longispina mucronatus
not present?
Meristella sp.

Ambocoelia umbonata
Devonochonetes coronatus
Eumetabolotoechia multicostum
Longispina mucronatus
Megastrophia concava
Meristella sp.

(continued)

Table 9.7
Continued.

MERISTELLA-CORAL BED Lower Mt. Marion (= Marcellus) Fm.	FALL BROOK CORAL BED Windom Member; Moscow Fm.

OTHER BRACHIOPODS:
  Mucrospirifer? consobrinus   *Mucrospirifer? consobrinus*
  Nucleospira cf. *N. concinna*   *Nucleospira concinna*
  not present?   *Parazyga hirsuta*
  not present?   *Pentamerella pavillionensis*
  Retichonetes cf. *R. vicinus*   *Retichonetes* sp.
  Schuchertella sp.   *Schuchertella* sp.
  Spinocyrtia cf. *S. granulosa*   *Spinocyrtia* cf. *S. granulosa*
  Tropidoleptus carinatus   *Tropidoleptus carinatus*

BRYOZOANS:
  Fenestella spp.   *Fenestella* spp.
  fistuliporoid mounds   fistuliporoid mounds
  Rhombopora sp.   *Rhombopora* sp.
  Sulcoretepora sp.   *Sulcoretepora* sp.

DOMINANT BIVALVES:
  Actinopteria decussata   *Actinopteria decussata*
  Cypricardella bellistriata   *Cypricardella bellistriata*
  Cypricardinia indenta   *Cypricardinia indenta*
  Goniophora sp.   *Goniophora* sp.
  Modiomorpha concentrica   *Modiomorpha concentrica*
  M. sp.   not present?
  Nuculites spp.   *Nuculites* spp.
  Nyassa arguata†   not present
  Palaeoneilo constricta   *Palaeoneilo constricta*
  P. emarginata   *P. emarginata*
  P. fecunda   *P. fecunda*
  Paracyclas lirata†   not present?
  Pseudaviculopecten sp.   *Pseudaviculopecten princeps*
  Pterinopecten undosus   *Pterinopecten undosus*

CRINOIDS:
  Ancyrocrinus bulbosus†   not present?
  Gennaeocrinus sp. (plates)   *Gennaeocrinus* sp. (plates)
  Gilbertsocrinus sp. (plates)   *Gilbertsocrinus* sp. (plates)

TRILOBITES:
  not present?   *Greenops* cf. *G. boothi*
  Phacops cf. *P. rana*   *Phacops rana*

TOTAL SPECIES RICHNESS (approx.)
  60–70 species   60–70 species

NOTE: Rank abundance of the most common species of corals and brachiopods are indicated by number after species name (modified from Brett, Miller, and Baird 1990).
  *new immigrant in Late Givetian.
  †Extinct before Late Givetian time.

species. Only five lineages out of more than fifty shown in the table fail to persist from the lowest to the highest samples. These may have become extinct. Overall, the background extinction value for the Hamilton appears from this compilation to be only about 10%, but it may actually be less than 1%. Only five species appearing in the upper coral bed samples are not present in the lower. Probably only one of these five is truly a new immigrant. The others are simply rare forms that have yet to be located in the relatively small examples of this specific lowest coral bed, as they are in fact present in many other lower Hamilton fossil-rich samples.

Even the dominance rankings of the most abundant eight taxa are similar. Note that the most abundant corals and brachiopods shown in table 9.7 are generally of the same rank in samples from both the Marcellus Formation and the Moscow Formation. Comparisons of other biofacies, such as the dysaerobic dark shale biofacies (Brett, Miller, and Baird 1990), reveal similar degrees of persistence of taxa.

The Onondaga-Hamilton faunal change is not merely a local phenomenon but appears to coincide with a global bioevent referred to as the Kacak or *Otamari* event (Walliser 1986, 1990; Boucot 1990b), which has been recognized by other workers in the Czech Republic, Morocco, Britain, Spain, and elsewhere (Chlupac and Kukal 1986; Truyols-Massoni et al. 1990; Boucot 1990b). This major change in the pelagic nowakiid-styliolinid faunas occurs near the Eifelian-Givetian boundary. According to Boucot (1988, 1990b), this event coincides rather precisely with the termination of cold water Malvinokaffric faunas in the Southern Hemisphere, and is thus associated with declining climatic gradients. It is also a time of very widespread transgression and deposition of anoxic black muds (Johnson, Klapper, and Sandberg 1985). In the Appalachian Basin the Kacak event is coincident with the second transgressive cycle of the Marcellus black shale. At this time we see the abrupt demise of the most typical Onondaga species and the influx of the Rhenish-Bohemian elements, suggesting major changes in water mass properties or newly available migration routes.

Similarly, the demise of the Hamilton-Tully fauna is associated with a second global event termed the *Pharciceras* or Taghanic event (House 1985; Boucot 1990b). This is a time of major extinction of ammonoids and other fauna, including the largest bryozoan extinction before the Permian (see Horowitz and Pachut 1993). Again, the faunal change is associated with the widespread deposition of black sha
Limestone (Johnson, Klapper, and Sandberg 1985).

associated with rises in eustatic sea level. Influx of anoxic waters and possible changes in climatic gradient appear to have caused extinction of long-ranging lineages (Boucot 1990a). However, major sea level high-stands may have also provided linkages between formerly isolated provinces, allowing for immigration of new taxa from outside the Appalachian Basin. For example, at the base of the Hamilton Group the appearance of the ammonoid *Agoniatites vanuxemi* records the influx of a normally Old World taxon.

Coordinated Stasis: Causes and Consequences

The patterns of distribution of faunal biofacies in the Silurian-Devonian interval of the Appalachian Basin suggest a new view of the dynamics of species-level change and stasis. In particular, we perceive a pervasive pattern of coordinated stasis.

Long-ranging associations of species or biofacies display very minor or no morphological change in most lineages of macrofossils. Very few taxa—fewer than 10%—become extinct, and few immigrant species appear during these time intervals. Within biofacies, at least 65% and sometimes more than 80% of morphospecies are found from lowest to highest samples of appropriate facies, although there may be a few outages (i.e., brief intervals wherein a normally common species is locally rare or absent) or substitutions in certain cycles.

Furthermore, overall patterns of species richness, rank abundance, and guild structure are relatively constant for a particular biofacies throughout an E-E subunit. We now recognize 14 such blocks of stability or coordinated stasis within Ecological-Evolutionary Unit V. These E-E subunits include at least four Frasnian faunas not discussed here. At present, we can only speculate as to the underlying causes of such pervasive stability.

It is likely that faunal associations or communities in many cases simply tracked stable, but laterally shifting, environments. Provided that the rate of environmental shift was not too rapid and that the particular environment continued to exist somewhere during an interval of time, most species appear to have been fully capable of migrating through periods of tens of thousands of years across distances of perhaps 300 kilometers, paralleling transgressions or regressions of paleoshorelines. Organisms seem to have followed their preferred environments.

Stabilizing selection in large populations certainly also plays a role in the morphologic stasis observed in so many species. However, communities appear to be resilient to change, including the introduction of new taxa. This is indicated by the fact that short-lived incursions of exotic faunal elements (incursion epiboles [Brett, Miller, and Baird 1990]) do not seem to have led to the establishment of permanent new members of tracking communities. Rather, these organisms appear in great abundance only in single cycles and then disappear abruptly, probably within a period of no more than a few hundred thousand years. There are relatively few instances within a block of stability, such as the Hamilton-Tully fauna, in which a faunal element appears partway through, becomes established, and then recurs in each successive cycle. This suggests that Appalachian Basin communities may have been relatively stable entities, even capable of resisting invasion by outside taxa. On the other hand, these minor epibole events may represent brief shifts in environmental conditions that temporarily allowed a species not normally viable in the Appalachian Basin to take hold. In any case, there are few instances of "exotic" taxa appearing within times of stability, thus suggesting that only a relatively minor influx of larvae from outside the system actually occurred. In general, it appears that only rapid major environmental changes were capable of knocking ecosystems out of equilibrium.

In contrast, during geologically brief intervals—probably no more than a half million years in duration—major faunal changes occurred in many biofacies more or less simultaneously. Certain biofacies were affected more than others. For example, nearshore siliciclastics display higher proportions of species carried over from one ecological-evolutionary subunit to the next, whereas carbonate shelf environments, especially reefs, show more dramatic change. Changes at some level, however, did occur in all biofacies. Such changes included the extinction of long-standing lineages. In general, only 10–30% of species crossed the boundary from one E-E subunit to the next. Furthermore, these brief intervals seem to have been times of greatly increased immigration of exotic species into the Appalachian Basin. This is particularly well illustrated by the Hamilton fauna, which appeared abruptly, with numerous Rhenish-Bohemian elements, very early in the Givetian Stage. In turn, new levels of species richness, dominance, and guild structure became established within biofacies geologically rapidly (within 100,000 to 500,000 years) and then remained very nearly constant throughout the several million years of coordinated stasis.

Again, we can speculate as to the cause of ecosystem breakdown and restructuring (also see Morris, Ivany, and Schopf 1992). There seems to be some relationship with major lowstands of sea level. Five of the Appalachian Basin faunal turnover events are coincident with sequence boundaries that represent lowstands. However, some major sequence boundaries occur within blocks of stability that seem to have had very little effect on tracking biotas. Perhaps only those sea level lowstands that completely eliminated shallow shelf habitats were effective in producing the terminal extinction and restructuring events.

An even stronger relationship exists between turnover events and major transgressions with widespread anoxia. Six of the major faunal turnovers, including the two boundaries of the Hamilton fauna, are closely associated with major sea level rise and deposition of black shales. Possibly the incursion of anoxic water into large areas of the basin may have had a major perturbing effect on biofacies and caused local extinction. There is also a possible association with climatic change in some cases. For example, evidence of global warming can be found at the Kacak event in the late Eifelian and perhaps of a slight further warming during the *Pharciceras* event, which terminated the Hamilton fauna (Boucot 1990a, 1990b).

In any case, the boundaries of most local Appalachian faunas are indeed associated with global bioevents. At least seven of the event boundaries that we recognize are also recognized by Boucot (1990a, 1990b) as minor global bioevents within the Silurian-Devonian interval. We thus conclude that environmental shifts—probably involving changing water masses and related to sea level and/or fluctuations in climates—were responsible ultimately for undermining the stability of co-occurring species. Major sea level rise also allowed the influx of new taxa along the newly opened migration routes.

Overall, our findings extend the concept of biomeres (introduced by Palmer in 1965) to times beyond the Cambrian. The pattern of persistent biotas, terminated by abrupt breakdown and restructuring events (Palmer 1965, 1984), is *not* unique to the Cambrian, nor to trilobites. The ecological-evolutionary subunits we have identified in the Silurian and Devonian are comparable to biomeres in most respects. Boucot (in press) also recognizes at least eight similar blocks of stability in the Ordovician. Furthermore, intervals of near-stasis in community structure, interrupted by restructuring events, are being recognized in other parts of the stratigraphic column, such as in Cenozoic mollusc faunas of the Atlantic Coastal Plain (W. Allmon, personal communication).

We suspect that blocks of stable faunas and large-scale punctuational patterns will come to be seen throughout the fossil record. Given the very limited amount of anagenetic change evident within the stable faunal blocks already identified, it seems likely that without the episodic perturbations and collapse of stable ecosystems there would have been little movement in the history of life.

Acknowledgments   Our manuscript was critically reviewed by Art Boucot, David Lehmann, Bob Anstey, Doug Erwin, and an anonymous reviewer. Their comments greatly improved the structure and contents of the finished work. Friedrich and Margaret Teichmann provided patient, skilled assistance with word processing; Wendy Taylor helped in figure preparation. Our research has been supported by NSF Grant EAR-9219807 and by a grant from the donors of the Petroleum Research Fund, American Chemical Society.

References

Baird, G. C. and C. E. Brett. 1983. Regional variation and paleontology of two coral beds in the Middle Devonian Hamilton Group of western New York. *Journal of Paleontology* 57:417–446.

Blodgett, R. B., D. M. Rohr, and A. J. Boucot. 1990. Early and Middle Devonian gastropod biogeography. In W. S. McKerrow and C. R. Scotese, eds., *Palaeozoic Palaeogeography and Biogeography. Geological Society Memoir* 12:277–284.

Boardman, R. S. 1960. Trepostomatous bryozoan of the Hamilton Group of New York State. *U.S. Geological Survey Professional Paper* 340:1–87.

Bofinger, J. M. and W. Compston. 1967. A reassessment of the age of the Hamilton Group, New York and Pennsylvania, and the role of inherited radiogenic [87]Sr. *Geochemica et Cosmochemica Acta* 31:2353–2359.

Bolton, T. E. 1957. Silurian stratigraphy and palaeontology of the Niagara Escarpment in Ontario. *Geological Survey of Canada Memoir* 289:1–145.

Boucot, A. J. 1975. *Evolution and Extinction Rate Controls.* Amsterdam: Elsevier.

———. 1982. Ecostratigraphic framework for the Lower Devonian of the North American Appalachian subprovince. *Neues Jahrbuch für Geologie und Paleontologie Abhandlungen* 163:81–121.

———. 1983. Does evolution take place in an ecological vacuum? II. *Journal of Paleontology* 57:1–30.

————. 1988. Devonian biogeography: An update. *Second International Symposium on the Devonian System. Canadian Society of Petroleum Geologists Memoir* 14(III):211–227.

————. 1990a. Phanerozoic extinctions: How similar are they to each other? In E. G. Kauffman and O. H. Walliser, eds., *Extinction Events in Earth History. Lecture Notes in Earth Sciences* 30:5–30. Berlin: Springer-Verlag.

————. 1990b. Silurian and pre–Upper Devonian bio-events. In E. G. Kauffman and O. H. Walliser, eds., *Extinction Events in Earth History. Lecture Notes in Earth Sciences* 30:125–132. Berlin: Springer-Verlag.

————. 1990c. Modern Paleontology: Using biostratigraphy to the utmost. *Revista Espanola de Paleontologia* 5:63–70.

————. 1995. The sporadic rather than periodic nature of extinction events. In *Extinction Events in Geological History: A Symposium*. Sociadad Mexicana de Paleontologia.

Brett, C. E., ed. 1986. Dynamic stratigraphy and depositional environments of the Hamilton Group (Middle Devonian) in New York State, Part I. *New York State Museum Bulletin* 457:1–157.

————. In press. Wenlockian fossil communities in New York State and adjacent areas. In A. J. Boucot and J. D. Lawson, eds., *Paleoecology of Silurian and Early Devonian Communities. Final Report of Project Ecostratigraphy*. Cambridge: Cambridge University Press.

Brett, C. E. and G. C. Baird. 1985. Carbonate shale cycles in the Middle Devonian of New York: An evaluation of models for the origin of limestones in terrigenous shelf sequences. *Geology* 13:324–327.

————. 1986. Symmetrical and upward-shallowing cycles in the Middle Devonian of New York State. *Paleoceanography* 1:431–445.

————. 1992. Coordinated stasis and evolutionary ecology of Silurian-Devonian marine biotas in the Appalachian Basin. *Geological Society of America Abstracts with Programs* 24:139.

Brett, C. E., W. M. Goodman, and S. T. LoDuca. 1990. Sequences, cycles, and basin dynamics in the Silurian of the Appalachian Basin. *Sedimentary Geology* 69:191–224.

Brett, C. E., K. B. Miller, and G. C. Baird. 1990. A temporal hierarchy of paleoecological processes in a Middle Devonian epeirical sea. In W. Miller, III, ed., *Paleocommunity Temporal Dynamics: The Long-Term Development of Multispecies Assemblages. Paleontological Society Special Publication* 5:178–209.

Brower, J. C. and O. B. Nye, Jr. 1991. Quantitative analysis of paleocommunities in the lower part of the Hamilton Group near Cazenovia, New York. In E. Landing and C. E. Brett, eds., *Dynamic Stratigraphy and Depositional Environments of the Hamilton Group (Middle Devonian) in New York State, Part II. New York State Museum Bulletin* 469:37–75.

Buehler, E. J. and I. H. Tesmer. 1963. Geology of Erie County, New York. *Buffalo Society of Natural Sciences Bulletin* 21:1–118.

Chlupác, I. and Z. Kukal. 1986. Reflection of possible global Devonian events in the Barrandian area C.S.S.R. In O. H. Walliser, ed., *Global Bio-events, a Crit-*

ical Approach. Lecture Notes in Earth Sciences 8:169–179. Berlin: Springer-Verlag.

Cleland, H. F. 1903. A study of the fauna of the Hamilton Formation of the Cayuga Lake section in central New York. *U.S. Geological Survey Bulletin* 206:13–112.

Eldredge, N. 1972. Systematics and evolution of *Phacops rana* (Green, 1832) and *Phacops iowensis* (Delo, 1935, trilobita) in the Middle Devonian of North America. *American Museum of Natural History Bulletin* 47:45–114.

Eldredge, N. and S. J. Gould. 1972. Punctuated equilibrium: An alternative to phyletic gradualism. In T. J. Schopf, ed., *Models in Paleobiology,* pp. 82–115. San Francisco: Freeman Cooper.

Feldman, H. R. 1980. Level-bottom brachiopod communities in the Middle Devonian of New York. *Lethaia* 13:27–46.

Gale, N. H. 1985. Calibration of the Palaeozoic time-scale; Ordovician, Silurian and Devonian periods. In N. J. Snelling, ed., *The Chronology of the Geological Record. The Geological Society Memoir* 10:81–88. Oxford: Blackwell.

Gillette, T. G. 1947. The Clinton of western and central New York. *New York State Museum Bulletin* 41:1–191.

Goldman, D. and C. E. Mitchell. 1990. Morphology, systematics, and evolution of Middle Devonian Ambocoeliidae (Brachiopoda), western New York. *Journal of Paleontology* 64:79–99.

Goldring, W. 1935. Geology of the Berne quadrangle. *New York State Museum Bulletin* 303:1–238.

———. 1943. Geology of the Coxsackie quadrangle, New York. *New York State Museum Bulletin* 332:1–274.

Gould, S. J. and N. Eldredge. 1977. Punctuated equilibria: The tempo and mode of evolution reconsidered. *Paleobiology* 3:115–151.

Grabau, A. W. 1906. Geology and paleontology of the Schoharie Valley. *New York State Museum Bulletin* 92:1–386.

Griffing, D. H. and C. A. Ver Straeten. 1991. Stratigraphy and depositional environments of the Marcellus Formation (Middle Devonian) in eastern New York. In J. R. Ebert, ed., *New York State Geological Association 63rd Annual Meeting Guidebook, Oneonta, New York,* pp. 205–249.

Harland, W. B., A. V. Cox, P. G. Llewellyn, C. A. G. Picton, A. G. Smith, and R. Walters. 1982. *A Geological Time Scale.* Cambridge: Cambridge University Press.

Harland, W. B., R. L. Armstrong, A. V. Cox, L. E. Craig, A. G. Smith, and D. G. Smith. 1990. *A Geological Time Scale, 1989.* Cambridge: Cambridge University Press.

Heckel, P. H. 1973. Nature, origin, and significance of the Tully Limestone, an anomalous unit in the Catskill delta, Devonian of New York. *Geological Society of America Special Paper* 138:1–244.

Horowitz, A. S. and J. F. Pachut. 1993. Specific, generic, and familial diversity of Devonian bryozoans. *Journal of Paleontology* 67:42–52.

House, M. R. 1985. Correlation of mid-Paleozoic ammonoid evolutionary events with global sedimentary perturbations. *Nature* 313:17–22.

Isaacson, P. E. and D. G. Perry. 1977. Biogeography and morphological conservatism of *Tropidoleptus* (Brachiopoda, Orthida) during the Devonian. *Journal of Paleontology* 51:1108–1122.

Johnson, J. G., G. Klapper, and C. A. Sandberg. 1985. Devonian eustatic fluctuations in Euramerica. *Geological Society of America Abstracts with Programs* 96:567–587.

Klapper, G. 1981. Review of New York Devonian conodont biostratigraphy. In W. A. Oliver Jr. and G. Klapper, eds., *Devonian Biostratigraphy of New York, Part I*, pp. 57–66. Washington, D.C.: International Union of Geology, Subcommission on Devonian Stratigraphy.

Koch, C. F. and J. P. Morgan. 1988. On the expected distribution of species' ranges. *Paleobiology* 14:126–138.

Koch, W. F., III. 1979. Brachiopod paleoecology, paleobiogeography, and biostratigraphy in the upper Middle Devonian of eastern North America: An ecofacies model for the Appalachian, Michigan, and Illinois Basins. Oregon State University, Ph.D. dissertation.

Landing, E. and C. E. Brett, eds. 1991. Dynamic stratigraphy and depositional environments of the Hamilton Group (Middle Devonian) in New York State, Part II. *New York State Museum Bulletin* 469:1–177.

Laporte, L. F. 1969. Recognition of a transgressive carbonate sequence within an epeiric sea: Helderberg Group (Lower Devonian) of New York State. *Society of Economic Paleontologists and Mineralogists Special Publication* 14:98–119.

Lieberman, B. S. 1993. Phylogeny of the Asteropyginae Delo, 1935, and the origin and evolution of the Middle Devonian Hamilton Group trilobite fauna. Columbia University, Ph.D. dissertation.

Lieberman, B. S., C. E. Brett, and N. Eldredge. In press. Patterns and processes of stasis in two species lineages from the Middle Devonian of New York State. *Paleobiology* 21.

McGhee, G. R. 1988. Evolutionary dynamics of the Frasnian-Famennian extinction event. In N. J. McMillan, A. Embry, and D. Glass, eds., *Proceedings of the Second International Symposium on the Devonian System. Canadian Society of Petroleum Geologists Memoir*, pp. 23–28.

———. 1990. The Frasnian-Famennian mass extinction record in the eastern United States. In E. G. Kauffman and O. H. Walliser, eds., *Extinction Events in Earth History. Lecture Notes in Earth Sciences* 30:161–168. Berlin: Springer-Verlag.

McIntosh, G. C. 1983. Crinoid and blastoid biogeography in the Middle Devonian (Givetian) of eastern North America. *Geological Society of America Abstracts with Programs* 15:171.

McKerrow, W. S., R. St. J. Lambert, and L. R. M. Cocks. 1985. The Ordovician, Silurian and Devonian periods. In N. J. Snelling, ed., *The Chronology of the*

Geological Record. The Geological Society Memoir 10:73–80. Oxford: Blackwell.

Morris, P. J., L. C. Ivany, and K. M. Schopf. 1992. Paleoecological stasis in evolutionary theory. *Geological Society of America Abstracts with Programs* 24:313.

Oliver, W. A., Jr. and J. E. Sorauf. 1981. Rugose coral biostratigraphy. In W. A. Oliver, Jr. and G. Klapper, eds., *Devonian Biostratigraphy of New York*, pp. 97–105. Washington, D. C.: Subcommission on Devonian Stratigraphy.

d'Orbigny, A. 1849–1852. *Cours Elementaire de Paleontologie et de Geologie Stratigraphique.* Paris: Masson.

d'Orbigny, A. 1850–1852. *Prodrome de Paleontologie.* Paris: Masson.

Palmer, A. R. 1965. Biomere—a new kind of biostratigraphic unit. *Journal of Paleontology* 39:149–153.

———. 1979. Biomere boundaries re-examined. *Alcheringa* 3:33–41.

———. 1984. The biomere problem: Evolution of an idea. *Journal of Paleontology* 58:599–611.

Pandolfi, J. M. and C. D. Burke. 1989a. Environmental distribution of long growth form in the favositid *Pleurodictum americanum. Lethaia* 22:69–84.

———. 1989b. Shape analysis of two sympatric coral species: Implications for taxonomy and evolution. *Lethaia* 22:183–193.

Roden, M. K., K. R. Parrish, and D. S. Miller. 1990. The absolute age of the Eifelian Tioga ash bed, Pennsylvania. *Journal of Geology* 98:282–285.

Scotese, C. R. 1986. *Phanerozoic Reconstructions: A New Look at the Assembly of Asia. University of Texas Institute of Geophysics Technical Report* 66.

Sheehan, P. M. 1985. Reefs are not so different—they follow the evolutionary pattern of level-bottom communities. *Geology* 13:46–49.

———. 1992. Patterns of synecology during the Phanerozoic. In E. C. Dudley, ed., *The Unity of Evolutionary Biology 1*, pp. 103–118. Portland: Discorides Press.

Sheldon, P. R. 1987. Parallel gradualistic evolution of Ordovician trilobites. *Nature* 330:561–563.

———. 1993. Making sense of microevolutionary patterns. In D. R. Lees and D. Edwards, eds., *Evolutionary Patterns and Processes. Linnean Society Symposium* 14:19–31.

Sorauf, J. E. and W. A. Oliver. 1976. Septal carinae and microstructures in Middle Devonian *Heliophyllum* (Rugosa) from New York State. *Journal of Paleontology* 50:331–343.

Stanley, S. M. 1979. *Macroevolution: Pattern and Process.* San Francisco: Freeman.

Swartz, C. K. 1913. Correlation of the Lower Devonian. In *Devonian, Lower*, pp. 96–132. *Maryland Geological Survey Special Publication.* Baltimore: Johns Hopkins Press.

Truyols-Massoni, M., R. Montesinos, J. L. Garcia-Alcade, and F. Leyva. 1990. The Kacak-Otomari event and its characterization in the Palentine Domain

(Cantabrian Zone, NW Spain). In E. G. Kauffman and O. H. Walliser, eds., *Extinction Events in Earth History. Lecture Notes in Earth History* 30:133–144. Berlin: Springer-Verlag.

Ver Straeten, C. A. 1993. Microstratigraphy and depositional environments of a Middle Devonian foreland basin: Berne and Otsego Members, Mount Marion Formation, eastern New York. In E. Landing, ed., *Studies in Stratigraphy and Paleontology in Honor of Donald W. Fisher. New York State Museum Bulletin* 481:367–380.

Vrba, E. S. 1985. Environment and evolution: Alternative causes of the temporal distribution of evolutionary events. *South African Journal of Science* 81:229–236.

Walliser, O. H. 1986. *Global Bio-events. Lecture Notes in Earth History* 8. Berlin: Springer-Verlag.

———. 1990. How to define global bioevents. In E. G. Kauffman and O. H. Walliser, eds., *Extinction Events in Earth History. Lecture Notes in Earth Sciences* 30:1–4. Berlin: Springer-Verlag.

Welch, D. 1991. Geographical variation and evolution in the middle Devonian brachiopod, *Mucrospirifer*. Virginia Polytechnical Institute and State University, Ph.D. dissertation.

Williamson, P. 1981. Paleontological documentation of speciation in Cenozoic molluscs from the Turkana Basin. *Nature* 293:437–443.

Zenger, D. H. 1965. Stratigraphy of the Lockport Formation (Middle Silurian) in New York State. *New York State Museum and Science Service Bulletin* 404:1–210.

———. 1971. Uppermost Clinton (Middle Silurian) stratigraphy and petrology, east central New York. *New York State Museum and Science Service Bulletin* 417:1–58.

10

Phylogenetic Trends and Speciation: Analyzing Macroevolutionary Processes and Levels of Selection

Bruce S. Lieberman

A hierarchical way of thinking, along with an emphasis on contingent explanations, has slowly crept into the field of evolutionary biology in recent years. This is a direct result of Eldredge and Gould's landmark 1972 paper, "Punctuated Equilibria: An Alternative to Phyletic Gradualism." In the paper they outlined a new way of looking at evolution in the fossil record which emphasized species, and this emphasis enhanced the contributions that paleontology could make to evolutionary theory. Here I consider the role of species and speciation in evolutionary theory in conjunction with a case history involving the turritellid gastropods, and with particular attention to trends as evolutionary phenomena.

The theoretical view of trends in evolution that has since become traditional was expounded by Dobzhansky (1951), Mayr (1942), and Simpson (1944, 1953). These three viewed major evolutionary patterns as originating at the level of changing allelic frequencies within a single population. In this way entire lineages evolve by moving through adaptive space (Simpson 1944; see Eldredge 1985 for a review).

The formulation of punctuated equilibria by Eldredge and Gould (1972) provides a new way of viewing trends and evolution in the fossil record, and their insights can best be elucidated by comparing their work with Simpson's. To begin, paleontologists Eldredge and Gould, along with Simpson, abnegate responsibility for formulating mechanisms of speciation, leaving such research to neontologists. In addition, in both Simpson's theory of quantum evolution and Eldredge and Gould's punctuated equilibria theory the presence of gaps in the fossil record play a prominent role.

The area of departure between Simpson and Eldredge and Gould centers around their differing ontology of species. To Simpson, species are just arbitrarily divided up chunks of the genealogical nexus (Simpson 1961; review in Wiley 1978 and Eldredge 1985). Simpson (1944) was not particularly concerned with the speciation process or species per se, but Eldredge and Gould (1972) recognized something unique about species that made them fundamental units of evolutionary change. To Eldredge and Gould (1972) species are real entities which can be recognized today in nature as well as through time. By viewing species as entities that persist for long periods of time without change and that also undergo episodes of rapid speciation relative to their total duration, Eldredge and Gould imbued species with unique properties. Species could be viewed as individuals, *sensu* Ghiselin (1974). They could also be viewed as creative entities or replicators (Dawkins 1976, 1982; Hull 1980), with speciation analogous to organismal birthing (Arnold and Fristrup 1982). Thus species can potentially serve as arenas for selection processes (Vrba 1989).

If species are not stable entities then they can evolve by the gradual transformation of phyletic lineages with the demarcation between species being essentially arbitrary. This ephemeral view of species can be contrasted strongly with the taxic perspective (Eldredge 1979) which stresses the role of species as entities that comprise the stuff of evolution rather than the ongoing process of evolution. A taxic approach engenders a view of evolutionary change in which the major features of evolution, trends, are the products of differential speciation (Gould 1990), with speciation not a process of transformation of lineages but rather of replication of lineages.

Punctuated equilibria is as much a theory about species as speciation. Once we view species as fundamental units of evolution, that is, as replicators which can be sorted and selected, then the data of paleontologists have important ramifications for evolutionary biology. In particular, our data base allows the detailed mapping out of trends.

The Relevance of Trends

A punctuated equilibria perspective is highly amenable to the analysis of trends because of the importance it ascribes to species. It treats the stuff of trends, species, as persistent entities, not arbitrarily divided up chunks of the genealogical nexus. Phylogenetic analysis can of course be performed on several entities, not just species, and trends can be viewed divorced from a punctuated equilibria ontology of species. However, when trends are placed in a context of species reality rather than evanescence, they are given far greater weight.

Trends are comprehended by a neo-Darwinian ontology of species, but their application at a larger level is limited. The neo-Darwinian perspective suggests that to deduce evolution all that is needed is the analysis of selection pressures in modern populations, followed by the simple extrapolation of those pressures over geologic time. Eldredge and Gould (1972) spurred tests for processes that could be involved in trends, and questioned the simple extrapolationist approach. The important role ascribed to species in their theory dictates that phylogeny should have a central role in paleontology and evolutionary biology, because it allows us to generate hypotheses about the evolution of groups of species.

When paleontologists have considered trends, they have chiefly been concerned with changes in numbers of taxa through time. Such changing diversity gradients have been the impetus for several studies. For example, Gould and Calloway (1980) considered the relationship between bivalve and brachiopod diversity. There have also been instances in which a group's diversification has been related to key morphological innovations. Such classes of trends differ from other trendlike phenomena. In these cases trends are being related explicitly to aptations (Gould and Vrba 1982), and simple numbers of taxa as a convenient descriptor of the trend will not suffice. Instead, the clade's diversification must be put in an actual evolutionary context.

An increase in the number of species may provide strong inferential support for a particular process. However, such taxonomic counts are no substitute for the stuff of evolutionary histories (Lieberman, Allmon, and Eldredge 1993). Species are the ground out of which trends grow; therefore, knowledge of the evolution of such should considerably elucidate our understanding of how trends proceed and what may be driving them.

The relationship of phylogenies to trends has become increasingly important in current evolutionary thought (e.g., Brooks and McLennan 1991; Farrell, Mitter, and Futuyma 1992). In order to elucidate this relationship, a case study of a trend is discussed herein. The trend is considered with regard both to different levels of selection and to ways of deducing the levels that may be responsible for the diversification of a clade. With the hierarchical expansion of evolutionary theory—given in the works of Eldredge and Gould (1972), Stanley (1979), Gould (1980), Eldredge and Salthe (1984), Vrba and Eldredge (1984), Eldredge (1985), and Buss (1987)—it was recognized that selection processes may operate at several levels of the genealogical hierarchy. These levels include the gene, the cell lineage, the population, and the species, as well as the organism. If we are to study trends in the search for mechanisms of evolutionary change we must consider several different hierarchical levels. A consideration of the relationship between levels of selection and trends can be instructive because it documents the testability of hierarchical theories of evolution.

Levels of Selection

One of the most acrimonious debates in evolutionary biology in recent years has been waged over the importance of selection at different hierarchical levels. Of course this presumes the existence of distinct hierarchical levels in nature, and the reader should refer to the work of Dawkins (1976, 1982), Gould (1980), Eldredge (1985), Eldredge and Salthe (1984), Vrba (1989), Vrba and Eldredge (1984), Eldredge (1989), and so on for various aspects of this debate. One of the levels of selection that has received considerable attention is species selection. Species selection has garnered increased exposure in the paleontological literature as a process implicated in the differential diversification of several clades relative to their sister taxa (Arnold and Fristrup 1982; Jablonski and Lutz 1983).

Introducing the Case Study: Turritellid Gastropods

The evolution of the gastropod family Turritellidae has been cited as one of the paradigm cases of species selection (Spiller 1977; Jablonski and Lutz 1983 [refs. therein]), and it is the focus here.

Fossil and extant turritellids, like most other gastropod species, have two larval types: planktonic and nonplanktonic (Allmon 1988, 1994; for alternative classificatory schemes for larval types see Jablonski and Lutz 1983 and Mileikovsky 1974). Planktonic larvae float in the water column for a few weeks before settling out and metamorphosing (Scheltema 1986). During this period, planktonic larvae may travel thousands of kilometers (Jablonski 1986; Hines 1986; Scheltema 1986). In contrast, nonplanktonic, lecithotrophic larvae develop in sessile egg sacs deposited by the parent and thus have poor dispersal capability. In the Neogene and extant turritellid biota, species with a nonplanktonic larval type outnumber species with a planktonic larval type by two or three to one (Spiller 1977; Allmon 1994).

The difference in the number of species with a nonplanktonic larval type relative to species with a planktonic larval type has been ascribed to species selection. The species selection claim is predicated on the supposition that nonplanktonic species are more prone to experience population subdivision due to their poor dispersal ability, which will lead first to incipient speciation and then to an elevated rate of diversification (Shuto 1974; Scheltema 1978; Jablonski 1986). Species with a planktonic larval type have good dispersal ability in their larval stage. This allows for gene flow between different populations within a species, and this tends to homogenize and buffer evolutionary change. Thus, such species lineages will be less likely to diversify. The acquisition of a nonplanktonic larval type and the increased speciation rate it engenders is held to be an emergent property of species, for the number of species with a nonplanktonic larval type increases not because of the greater fitness of their component organisms but because of their increased propensity to speciate.

In order to examine one of the paradigm cases of species selection, I, along with my collaborators in Lieberman, Allmon, and Eldredge 1993, performed a phylogenetic analysis of the turritellids to assess the evolution of larval types in this family. One can view cases of species selection just as one views the phenomenon of organismal adaptation, but as an adaptation at the species level (Arnold and Fristrup 1982). Adaptation relates to differential survival and differential reproduction (Gould and Vrba 1982; Coddington 1988; Hull 1980; Eldredge 1985). With a hierarchical expansion to the species level, such an adaptation may impart either greater longevity to species (group selection as portrayed in Wynne-Edwards 1962), or greater propensity to speciate (group selection as portrayed in Wright 1931). However, some authors have suggested that species longevity is not a species-level character, as it is controlled by the survival of component organisms, and thus it is not an appropriate criterion for recognizing the operation of species selection (Vrba 1984a; Eldredge 1989). The replicative nature of species is held to be an emergent property of species following Eldredge (1985, 1989), Vrba (1989), and Lieberman (1992).

Potential Mechanisms Driving Turritellid Gastropod Evolution

The practice of testing hypothetical processes with phylogenies was first propounded by Eldredge and Cracraft (1980). Since then, interest has burgeoned in the relationship between evolutionary histories and processes (Brooks and McLennan 1991; Farrell, Mitter, and Futuyma 1992). If we are to consider the hypothesis of species selection as a theory within the realm of the testable, that is, a part of science, then we should assiduously search for a means of verification. Other mechanisms do exist which may be relevant to the trend in turritellids toward a greater number of species with a nonplanktonic larval type relative to species with a planktonic larval type, and these need to be briefly outlined.

One mechanism that may drive the trend is organismal adaptation mediated by constraints related to the loss of feeding structures in the larval stage (Strathmann 1978a,b; Hansen 1982; Raff 1987; Wray and Raff 1991). Under this mechanism, the number of species with a nonplanktonic larval type increases because of selection gradients. The reacquisition of a long larval stage in organisms that lacked the structures needed for larval feeding would be selected against, as such organisms would starve. Thus, due to selection at the organismal level, species would retain a short nonplanktonic larval type. Planktonic lineages could presumably give rise to both planktonic and nonplanktonic species (i.e., long-lived and short-lived larvae).

Another mechanism possibly driving the trend toward more nonplanktonic species relies on selection at the level of different cell lineages (Lieberman, Allmon, and Eldredge 1993). This mechanism is based on the fact that the chief differences between organisms with planktonic and nonplanktonic larval types are differences in developmental timing and reproductive maturation (Blackwell and Ansell 1974; Fretter and Graham 1962; Lebour 1937; Perron 1981, 1986; Strathmann 1978b, 1986; Thorson 1946). Metamorphosis from veliger larva to adult in gastropods corresponds to the period in development when planktonic larvae become competent and settle out of the water column to the bottom—or approximately to the time when nonplanktonic larvae hatch from their eggs. The torsional process is associated with this period of metamorphosis in both larval types (D'Asaro 1965; Fretter and Graham 1962).

Planktonic species of gastropods have an extended veliger larval life (Scheltema 1986; Thorson 1950). By contrast, nonplanktonic larvae (of mesogastropods) typically metamorphose and reach adulthood relatively rapidly, and thus have a relatively short veliger larval life (Lebour 1933, 1937; Thorson 1946; Blackwell and Ansell 1974; Fretter and Graham 1962). These singular features have important implications when considered in light of embryological analyses of early gastropod development. (Here nonplanktonic larvae refers only to nonbrooded nonplanktonic larvae, as no brooded turritellids were used in this phylogenetic analysis. The development of brooded larvae may be very different from nonbrooded nonplanktonic larvae.)

Buss (1987, 1988) presented a detailed overview and consideration of embryological studies of the differentiation of the primordial germ cells in several phyla. He recognized that during organismal development there may be a delay between maternal differentiation and the timing of germline sequestration. Such a delay can be viewed as a window in development. As long as this window remains open, alterations in ontogeny can be inherited by future generations because the germ-line cells have not yet been sequestered from the somatic cells (Buss 1987).

Embryological studies suggest that in molluscs there is variation in the length of the delay between maternal predetermination and germ-line sequestration (Dohmen 1983; Moor 1983). Thus, the length of time in which the window of heritable variation remains open ranges from very short in some gastropod species to quite long in others. However, it must be noted that these studies need to be bolstered, first by increasing the number of species analyzed, as thus far embryological preparations have been performed on relatively few, albeit phylogenetically disjunct, species of gastropods. More important, embryological preparations that suggest the existence of a window may mask earlier determination at the molecular level (Buss 1987, 1988). Additional studies using molecular markers, therefore, must be conducted.

Embryological evidence from histological preparations of gastropods suggest that the germ-line is sequestered during metamorphosis (Conklin 1897; Raven 1958; Fretter and Graham 1962; D'Asaro 1965; Verdonk and Biggelaar 1983). Germ-line sequestration would act as a developmental mechanism driving the differential diversification of species with a nonplanktonic larval type in the following manner. Larval style appears related to the timing of metamorphosis; on the basis of histological preparations metamorphosis, and thus larval style, is linked to the timing of germ-line sequestration. A long larval life, as epitomized by organisms in species with a planktonic larval type, implies a long period of time before metamorphosis transpires and thus before the germ-line is sequestered. In such organisms a relatively long period of time will transpire during which heritable developmental innovations can arise. By contrast, the relatively rapid attainment of metamorphosis in gastropod species with a nonplanktonic larval type creates a narrow window during which heritable variations in ontogeny can occur (Lieberman, Allmon, and Eldredge 1993).

Developmental changes, as a first approximation, can be assumed to have an equal probability of either lengthening or shortening larval life. Once larval life is shortened, however, the size of the window for heritable changes in development is decreased. As the duration of this developmental window decreases, developmental modifications will become increasingly less likely to occur (Lieberman, Allmon, and Eldredge 1993). Thus, once a species with a short nonplanktonic larval period originates, developmental mode will probably be conserved, simply because there is less opportunity for developmental style to change. Thus, driven by a developmental mechanism constraining future opportunities for change in phylogeny, the number of species with a nonplanktonic larval type should increase—even assuming no increased propensity for speciation in nonplanktonic lineages. Species lineages with a nonplanktonic larval type would be expected to give rise almost exclusively to species with a nonplanktonic larval type, whereas species lineages with a planktonic larval type could give rise to both planktonic and nonplanktonic species. The biases produced by this developmental mechanism would operate in a way analogous to the biases produced by meiotic and molecular drive (Lewontin and Dunn 1960; Dover 1982). I thus name this mechanism *cell-lineage drive.*

Overall, any one of three levels of the genealogical hierarchy—species, organisms, and cell lineages—could be the locus of selection that is, in turn, responsible for motivating the trend at issue here: differential diversification between gastropods with planktonic and nonplanktonic larval forms. Species selection, for example, may be driving differential diversification by producing an elevated speciation rate in nonplanktonic lineages. Or, selection at the organism level may be driving species to their optimally adaptive developmental modes and rates. Finally, selection among cell lineages acting during development may bias the openness of future generations to heritable changes in ontogeny, producing a developmentally rigid, nonplanktonic lineage which can give rise predominantly to nonplanktonic species. Planktonic species lineages, meanwhile, will be able to produce both planktonic and nonplanktonic species.

Here I will explore the hypothesis that species selection (selection at the species level) is actually driving speciation in turritellid gastropods. The other two processes act as overlays, constraining the direction of ongoing speciation.

Choosing Among Different Levels of Selection

We can discuss the evolutionary patterns we see in nature, and we can formulate hypotheses about the processes that may have produced these patterns, for, "the most important connection between (pattern and process) . . . involves the comparison of both intrinsic and extrinsic features of organisms predicted from theories of process, with those actually found in nature" (Eldredge and Cracraft 1980, p. 4). Vrba (1984b, 1990) demonstrated how to tease apart the relationship between pattern and the processes acting at different levels of the genealogical hierarchy. She recognized that the phenomenological levels of organismal proliferation—engendered by organismal adaptation and taxic proliferation, mediated by the effect hypothesis—can be totally disjunct. In terms of total numbers of organisms, impalas and alcelaphines are roughly equivalent; however, the number of alcelaphine species greatly exceeds the number of impala species.

Phylogenetic analysis can be used as an arbiter of hypotheses about the evolutionary processes, but a phylogeny alone cannot refute the validity of a particular mechanism. Instead, it can tell us if that mechanism was important for a particular clade's diversification. Phylogenetic models assuming only one of the mechanisms was operating can be constructed and then compared to an actual phylogeny of the turritellids that displays the evolution of larval type in that family. Earlier tests of species selection in the Turritellidae (e.g., Spiller 1977) were circular, for they used the very character to define the species-rich nonplanktonic clade, a nonplanktonic larval type, that was putatively responsible for driving the diversification of that clade.

To avoid such circularity, I have used DNA sequences to reconstruct the phylogeny of the turritellids. DNA sequences provide a means of testing evolution independent of developmental and morphological traits that play a prominent role in the trend being studied. The use of DNA necessarily limited my study (and that of Lieberman, Allmon, and Eldredge 1993) to extant turritellids, but the patterns motivating both were discovered in the fossil record.

In this study the first phylogenetic model assumed species selection was the only process driving the differential diversification of nonplanktonic species. This process predicts that the acquisition of a particular feature, emergent at the species level, is an adaptation or key innovation that facilitates the proliferation of species within a given clade. The origin of a nonplanktonic larval type is held to be the adaptation driving the trend in the turritellid gastropods (Lieberman, Allmon, and Eldredge 1993).

The demonstration of selection pressure is the essence of the search for adaptation. However, after demonstrating selection, epistemologies of adaptation tend to diverge. Some authors (e.g., Coddington 1988) have proposed that for a character to be an organismal adaptation it must be able to define a monophyletic group. This reasoning receives some support by means of an example. "Wings" have been acquired at least three times in tetrapod history: in birds, bats, and pterosaurs. Wings cannot be viewed as an adaptation of tetrapods because the three distinct categories of morphologies that achieve flight are not homologous. They are morphological traits separated by evolutionary connectivity. The state of wingedness could be viewed as a functional property present in all organisms that do fly. However, the attendant adaptations would have to be recognized in a purely economic and not genealogic context. This exercise demonstrates that the concept of adaptation could be severed from the constraints of monophyletic ancestor-descendant relationships (Harvey and Pagel 1991).

It should be clear that there is a difference between organismal adaptation from the standpoint of synapomorphy and organismal adaptation from the standpoint of parallelism. This duality is analogous to the appreciation of homology as a strictly evolutionary concept (e.g., Eldredge and Cracraft 1980; Gould and Vrba 1982; Coddington 1988), as a mechanistic concept (Wagner 1989), or as something delineated by emergence at particular hierarchical levels (Striedter and Northcutt 1991). However, the chief evidence for adaptation at the *species level* is information about clade shape and topology (Vrba 1989). Thus, the criterion for recognizing adaptations at the species level must be placed in the context of clades, which are monophyletic lineages comprising an ancestor and all of its descendants.

In this paper, I will therefore view species-level adaptations as Coddington (1988) views them. For a character to be an adaptation at the species level it must define a monophyletic group. Without a single acquisition of a nonplanktonic larval type in the Turritellidae, we can reject species selection as the sole cause of the proliferation of species with a nonplanktonic larval type. This is because there would not have been a single adaptation in the turritellids for increased diversification of nonplanktonic species.

The hypothesis of species selection surely predicts that the clade marked by the acquisition of a nonplanktonic larval type will be more diverse than its sister taxon, even though both clades, as sister taxa, must have the same age. I have used the predicted evolutionary results of a case of species selection to formulate my phylogenetic model #1. The tree topology of model #1 is shown in figure 10.1. If all speciation events occurred with regularity, each branch on the tree lying within a monophyletic nonplanktonic clade should have a series of bifurcations. However, the assumption that all speciation events should occur with equal probability in the nonplanktonic clade would produce a test for species selection that is too rigorous and too confining. Instead, a generally increased tendency to diversify, resulting in an asymmetrical phylogeny, is all that is required to confirm the hypothesis. Diversification within each of the "nonplanktonic" branches of the pectinate cladogram shown in figure 10.1 would provide additional support for species selection.

Model #2, shown in figure 10.2, is a hypothetical phylogeny based on the predictions of Strathmann's (1978a,b) and Wray and Raff's (1991) organismal adaptation hypothesis and on the developmental mechanism of cell-lineage drive propounded here. These processes, operating at two different genealogical levels, seem to have similar phylogenetic implications. Evidence for either of these processes would be multiple or homoplastic originations of nonplanktonic species from *different* planktonic lineages. In addition, nonplanktonics should rarely, if ever, give rise to planktonics. This is because both mechanisms predict that it will be difficult to reacquire a planktonic larval style once it has been lost (Lieberman, Allmon, and Eldredge 1993).

As in model #1, the phylogenetic predictions of model #2 are strongly affected by one's views of adaptation and homology. Model #2 predicts multiple acquisitions of similar emergent properties of development, such

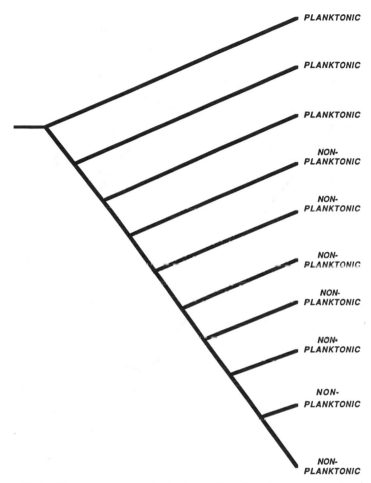

Figure 10.1 Phylogenetic model #1, the predicted evolutionary results if species selection were driving the differential diversification of species with a non-planktonic larval type. This model serves as a test for the operation of species selection. Depicted is the hypothetical fit of larval characters to a tree in which proliferation is associated with the singular acquisition of a nonplanktonic larvae. Diversification is hypothesized to have been produced by incipient speciation, facilitated by the population structure of species of sessile organisms possessing a nonplanktonic larval stage.

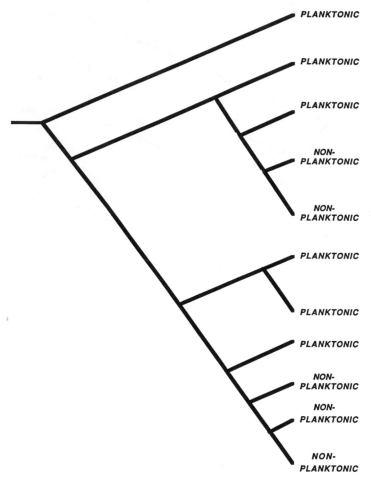

Figure 10.2 The phylogeny predicted by model #2. It serves as a test for both the developmental mechanism and organismal adaptation mediated by developmental constraints. Shown are the distribution of planktonic and nonplanktonic larval types in a phylogeny predicted to have been directed by these mechanisms. Owing to developmental mechanisms and/or organismal adaptation, planktonic larvae are likely to give rise to nonplanktonic larvae, but the reverse is not true. The switch from a planktonic to a nonplanktonic larval type occurs in different planktonic lineages.

as rate, mode, and larval style, along with a reduced or absent tendency to revert to a planktonic larval type. If developmental processes are hypothesized to be responsible for this pattern then we can view these multiple acquisitions of a nonplanktonic larval type as homologies *sensu* Wagner (1989) because they involve the manifestation of similar processes and mechanisms of development. By the same token, and in analogy to part of the argument of Striedter and Northcutt (1991), a nonplanktonic larval type is a homology at the level at which it is emergent in the genealogical hierarchy—the level of development. Therefore, the independent acquisition of similar emergent processes of development could be viewed as homologous traits in the parlance of Wagner (1989) and in part of Striedter and Northcutt (1991). In both cases, demonstration of "homology" of mechanism or of "homology" of emergent traits at a particular hierarchical level suggests that we should recognize selection and adaptation occurring at the level of cell lineages. If organismal adaptation is the process responsible for the phylogenetic pattern manifested in model #2, then we can view the acquisition of a nonplanktonic larval type as an adaptation in the sense of Harvey and Pagel (1991).

Results of Phylogenetic Analysis

The phylogeny of the turritellids we constructed used a parsimony analysis employing transversions, additions, and deletions for 300 bases of the 16s ribosomal RNA gene in the mitochondria. A single most parsimonious tree of length 114 steps, consistency index 0.57 (once autapomorphies were removed), and retention index 0.47 was generated by the exhaustive search options in Hennig86 (Farris 1988) and PAUP 3.0q (Swofford 1990). This tree is shown in figure 10.3. Larval type is mapped onto this tree in figure 10.4. Detailed discussions of tree construction, sequence alignment, and larval type assessment are given in Lieberman, Allmon, and Eldredge 1993.

The phylogeny suggests that a planktonic larval type is primitive for the turritellids, which conforms with the conclusions of morphological analyses done on fossil taxa by Allmon (1994). In addition, it appears that a nonplanktonic larval type has evolved at least twice, if one assumes an autapomorphic reversion to planktonic in *Turritella gonostoma*. A cladogram for the turritellids in which a nonplanktonic larval type maps

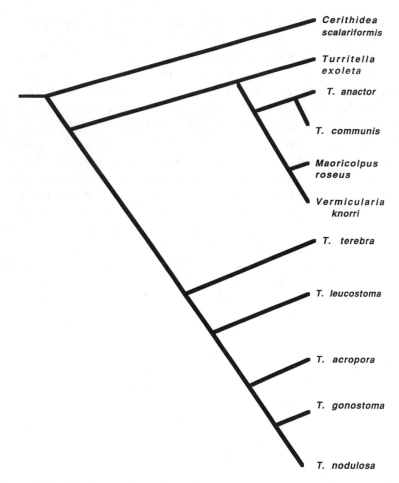

*Cerithidea
scalariformis*

*Turritella
exoleta*

T. anactor

T. communis

*Maoricolpus
roseus*

*Vermicularia
knorri*

T. terebra

T. leucostoma

T. acropora

T. gonostoma

T. nodulosa

Figure 10.3 The shortest length tree for the turritellid gastropods produced by phylogenetic analysis of 300 base pairs of 16S ribosomal mitochondrial DNA. The tree length is 114 steps, the consistency index is 0.57, and the retention index is 0.47.

as monophyletic requires at least twelve more steps than does the single most parsimonious tree shown in figure 10.4 (Lieberman, Allmon, and Eldredge 1993).

Our phylogenetic work on the evolution of larval type in the turritellids suggests that (1) acquisition of a nonplanktonic larval type cannot be viewed as a single adaptation that leads to the diversification of the turritellids, and (2) the repeated acquisition of similar emergent properties of development—such as rate, mode, and larval style, along with the reduced or absent tendency to revert to a planktonic larval type—suggests that developmental processes play a role in this trend. Based on the models formulated here, species selection alone cannot explain the differential diversification of nonplanktonic species relative to species with a planktonic larval type.

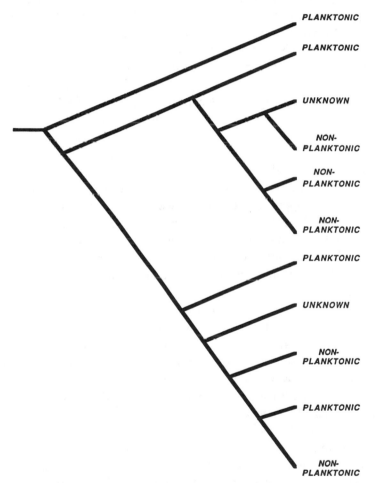

Figure 10.4 The same phylogeny as figure 10.3, but showing the larval type of each species. A planktonic larval type is primitive for the turritellids, and a nonplanktonic larval mode is acquired at least twice.

Reductionism and Assessing Causality

Our phylogenetic work thus suggests that cell-lineage drive, organismal adaptation, or both (hence, selection acting at two levels of the genealogical hierarchy) could explain the trend we studied in the Turritellidae. That is, assuming (1) speciation rates for planktonic and nonplanktonic lineages are equal, (2) a nonplanktonic lineage gives rise to a nonplanktonic species 100% of the time, and (3) a planktonic lineage gives rise to a nonplanktonic species only 50% of the time, then it would be quite easy to arrive at a two- or three-to-one ratio of nonplanktonic species to planktonic species. This is in fact the value known from the fossil Neogene and extant turritellid biota (Spiller 1977; Allmon 1994). Williams (1966, p. 4) has suggested that selective forces "should be attributed to no higher a level of organization than is demanded by the evidence." This dictum would lead one to conclude that cell-lineage drive is the force behind turritellid evolution.

Such a conclusion is reductionist in tone. Indeed, the entire analysis of the importance of selective forces at different levels of the genealogical hierarchy can, in a way, be viewed as predicated on the tenets of reductionism. The properties of the whole would devolve into properties of parts (Rosen 1991). However, one cannot "solve a 3-body problem by breaking a 3-body system . . . into 3 one-body systems or even a 2-body and [a] 1-body system" (Rosen 1991, p. 22). "In hierarchical systems, the upper level gives meaning to the level of focus, the lower level" (Rosen 1991, p. xi).

The different levels are emphasized and dissected by those interested in the applications of hierarchy theory to evolutionary biology in order to illustrate the wealth of possible mechanisms involved, not to dignify a single level as the sole shaper of an evolutionary trend. But, the appeal of pluralism can be lost when demands for testable hypotheses emerge. The models I presented earlier in this paper were introduced to test the relevance of particular mechanisms. The notion that in twentieth century science "the law of causality . . . is a relic of a bygone age, surviving, like the monarchy, only because it is erroneously supposed to do no harm" (Bertrand Russell) is rejected. Causal processes can be recognized, but as in the nature versus nurture debate in psychology, it appears that a bit of both are important. Naive reductionism cannot be substituted for the assiduous attempt to discern a series of causes, each with different degrees of importance.

Clearly, more than just a phylogeny is needed to refute a particular hypothesized process. Although our own results show no evidence of species selection, this work alone cannot foreclose the possibility that species selection could in some way be contributing to the diversification of the two lineages that independently acquire a nonplanktonic larval type in the turritellids. To tease apart the contributions of selection at particular levels of the genealogical hierarchy, we will need detailed analyses of the fossil record to discern actual differences in the speciation rates of non-planktonic lineages relative to planktonic lineages. We will also need a much better understanding of the relationship between fitness, larval period, and larval feeding. Finally, we will need molecular markers of development to assess what differences, if any, exist in the timing of germ-line sequestration in nonplanktonic versus planktonic species of gastropods. If the window in which the germ-line is not sequestered is, as we have speculated, longer in planktonic larvae than in nonplanktonic larvae we can use actual differences in timing deduced from molecular studies to construct computer models that would show what biases in diversity through time we would expect if only cell-lineage drive were operating. It is safe to say now that species selection is not the sole explanation for the documented trend in the turritellids.

Overall, this analysis presented a way of considering hypotheses generated from the study of fossils and extant organisms in conjunction with knowledge of organismal development in order to assess forces driving trends. By combining data from several different fields one naturally considers more focal levels in the genealogical hierarchy, and the results will be more comprehensive than those based on a single data base. Trends constitute the major patterns we see in the fossil record, and macroevolutionary studies that employ phylogenetic analyses provide a technique for generating detailed maps of trends. The tangible evidence of stasis and the concomitant presence of distinct species in the fossil record, as recognized by Eldredge and Gould (1972), is the single greatest justification for macroevolutionary studies because together they demonstrate that modern biological theories must be held accountable to macroevolutionary patterns.

Acknowledgments   I thank Niles Eldredge and Douglas Erwin for their comments on the manuscript. I also thank Warren Allmon for his discussions and collaboration.

References

Allmon, W. D. 1988. Ecology of recent turritelline gastropods (Prosobranchia, Turritellidae): Current knowledge and paleontological implications. *Palaios* 3:259–284.

———. 1994 (in press). Systematics and evolution of Cenozoic American Turritellidae. I. Paleocene and Eocene species from the U.S. Gulf and Atlantic coastal plains related to "*Turritella mortoni* Conrad" and "*Turritella humerosa* Conrad." *Paleontigraphica Americana.*

Arnold, A. J. and K. Fristrup. 1982. The theory of evolution by natural selection: A hierarchical expansion. *Paleobiology* 8:113–129.

Blackwell, W. M. and A. D. Ansell. 1974. The direct development of the bivalve *Thyasira gouldi. Thalassia Jugoslavica* 10:23–43.

Brooks, D. R. and D. A. McLennan. 1991. *Phylogeny, Ecology, and Behavior.* Chicago: University of Chicago Press.

Buss, L. W. 1987. *The Evolution of Individuality.* Princeton: Princeton University Press.

———. 1988. Diversification and germ-line determination. *Paleobiology* 14:313–321.

Coddington, J. A. 1988. Cladistic tests of adaptational hypotheses. *Cladistics* 4:3–22.

Conklin, E. G. 1897. The embryology of *Crepidula. Journal of Morphology* 13:1–226.

D'Asaro, C. N. 1965. Organogenesis, development, and metamorphosis in the queen conch *Strombus gigas,* with notes on breeding habits. *Bulletin of Marine Science* 15:359–416.

Dawkins, R. 1976. *The Selfish Gene.* New York: Oxford University Press.

———. 1982. *The Extended Phenotype.* San Francisco: Freeman.

Dobzhansky, T. 1951. *Genetics and the Origin of Species.* 3rd ed. New York: Columbia University Press.

Dohmen, M. R. 1983. Gametogenesis. In N. H. Verdonk, J. A. M. van den Biggelaar, and A. S. Tompa, eds., *The Mollusca, Volume 3, Development,* pp. 1–48. New York: Academic Press.

Dover, G. A. 1982. Molecular drive: A cohesive mode of species formation. *Nature* 299:111–117.

Eldredge, N. 1979. Alternative approaches to evolutionary theory. In J. H. Schwartz and H. B. Rollins, eds., *Models and Methodologies in Evolutionary Theory. Bulletin of the Carnegie Museum of Natural History* 13:7–19.

———. 1985. *Unfinished Synthesis.* New York: Oxford University Press.

————. 1989. *Macroevolutionary Dynamics*. New York: McGraw-Hill.

Eldredge, N. and J. Cracraft. 1980. *Phylogenetic Patterns and the Evolutionary Process*. New York: Columbia University Press.

Eldredge, N. and S. J. Gould. 1972. Punctuated equilibria: An alternative to phyletic gradualism. In T. J. M. Schopf, ed., *Models in Paleobiology*, pp. 82–115. San Francisco: Freeman, Cooper.

Eldredge, N. and S. Salthe. 1984. Hierarchy and evolution. *Oxford Surveys in Evolutionary Biology* 1:182–206.

Farrell, B. D., C. Mitter, and D. J. Futuyma. 1992. Diversification at the insect-plant interface. *BioScience* 42:34–42.

Farris, J. S. 1988. Hennig86. Version 1.5.

Fretter, V. and A. Graham. 1962. *British Prosobranch Molluscs*. London: Ray Society.

Ghiselin, M. T. 1974. A radical solution to the species problem. *Systematic Zoology* 23:536–544.

Gould, S. J. 1980. Is a new and general theory of evolution emerging? *Paleobiology* 6.119 130.

————. 1990. Speciation and sorting as the source of evolutionary trends, or "things are seldom what they seem." In K. J. McNamara, ed., *Evolutionary Trends*. London: Belhaven.

Gould, S. J. and C. B. Calloway. 1980. Clams and brachiopods—ships that pass in the night. *Paleobiology* 6:383–396.

Gould, S. J. and E. S. Vrba. 1982. Exaptation—a missing term in the science of form. *Paleobiology* 8:4–15.

Hansen, T. 1982. Modes of larval development in Early Tertiary neogastropods. *Paleobiology* 8:367–377.

Harvey, P. H. and M. D. Pagel. 1991. *The Comparative Method in Evolutionary Biology*. New York: Oxford University Press.

Hines, A. H. 1986. Larval problems and perspectives in the histories of marine invertebrates. *Bulletin of Marine Science* 39:506–525.

Hull, D. L. 1980. Individuality and selection. *Annual Review of Ecology and Systematics* 11:311–332.

Jablonski, D. 1986. Larval ecology and macroevolution in marine invertebrates. *Bulletin of Marine Science* 39:565–587.

Jablonski, D. and R. A. Lutz. 1983. Larval ecology of marine benthic invertebrates: Paleobiological implications. *Biological Reviews* 58:21–89.

Lebour, M. V. 1933. The eggs and larvae of *Turritella communis* Lamarck and *Aporrhais pes-pelicani* (L.). *Journal of the Marine Biological Association of the United Kingdom* 18:499–506.

————. 1937. The eggs and larvae of the British prosobranchs with special reference to those living in the plankton. *Journal of the Marine Biological Association of the United Kingdom* 22:105–166.

Lewontin, R. and L. C. Dunn. 1960. The evolutionary dynamics of a polymorphism in the house mouse. *Genetics* 45:705–722.

Lieberman, B. S. 1992. An extension of the SMRS concept into a phylogenetic context. *Evolutionary Theory* 10:157–161.

Lieberman, B. S., W. D. Allmon, and N. Eldredge. 1993. Levels of selection and macroevolutionary patterns in the turritellid gastropods. *Paleobiology* 19:205–215.

Mayr, E. 1942. *Systematics and the Origin of Species.* New York: Dover.

Mileikovsky, S. A. 1974. Types of larval development in marine bottom invertebrates: An integrated ecological scheme. *Thalassia Jugoslavica* 10:171–179.

Moor, B. 1983. Organogenesis. In N. H. Verdonk, J. A. M. Biggelaar, and A. S. Tompa, eds., *The Mollusca, Volume 3, Development,* pp. 123–178. New York: Academic Press.

Perron, F. E. 1981. Larval growth and metamorphosis of *Conus* (Gastropoda: Toxoglossa). *Pacific Science* 35:25–38.

———. 1986. Life history consequences of differences in developmental mode among gastropods in the genus *Conus. Bulletin of Marine Science* 39:485–497.

Raff, R. A. 1987. Constraint, flexibility, and phylogenetic history in the evolution of direct development in sea urchins. *Developmental Biology* 119:6–19.

Raven, C. P. 1958. *Morphogenesis: The Analysis of Molluscan Development.* New York: Pergamon.

Rosen, R. 1991. *Life Itself.* New York: Columbia University Press.

Scheltema, R. S. 1978. On the relationship between dispersal of pelagic veliger larvae and the evolution of marine prosobranch gastropods. In B. Battaglia and J. A. Beardmore, eds., *Marine Organisms: Genetics, Ecology, and Evolution,* pp. 303–322. New York: Plenum.

———. 1986. On dispersal and planktonic larvae of benthic invertebrates: An eclectic overview and summary of problems. *Bulletin of Marine Science* 39:290–322.

Shuto, T. 1974. Larval ecology of prosobranch gastropods and its bearing on biogeography and paleontology. *Lethaia* 7:239–256.

Simpson, G. G. 1944. *Tempo and Mode In Evolution.* New York: Columbia University Press.

———. 1953. *The Major Features of Evolution.* New York: Columbia University Press.

———. 1961. *Principles of Animal Taxonomy.* New York: Columbia University Press.

Spiller, J. 1977. Evolution of turritellid gastropods from the Miocene and Pliocene of the Atlantic Coastal Plain. State University of New York at Stony Brook, Ph.D. dissertation.

Stanley, S. M. 1979. *Macroevolution.* San Francisco: Freeman.

Strathmann, R. R. 1978a. The evolution and loss of feeding larval stages of marine invertebrates. *Evolution* 32:894–906.

———. 1978b. Progressive vacating of adaptive types during the Phanerozoic. *Evolution* 32:907–914.

————. 1986. What controls the type of larval development? Summary statement for the evolution session. *Bulletin of Marine Science* 39:616–622.

Striedter, G. F. and R. G. Northcutt. 1991. Biological hierarchies and the concept of homology. *Brain, Behavior and Evolution* 38:177–189.

Swofford, D. 1990. PAUP: Phylogenetic analysis using parsimony, Version 3.0q. Champaign: Illinois Natural History Survey.

Thorson, G. 1946. Reproduction and larval development of Danish marine bottom invertebrates. *Meddelelser fra Kommissionen for Danmarks Fiskeri-og Havundersogelser Serie, Plankton* 4:1–523.

————. 1950. Reproductive and larval ecology of marine bottom invertebrates. *Biological Reviews* 25:1–45.

Verdonk, N. H. and J. A. M. van den Biggelaar. 1983. Early development and the formation of the germ layers. In N. H. Verdonk, J. A. M. van den Biggelaar, and A. S. Tompa, eds., *The Mollusca, Volume 3, Development*, pp. 91–122. New York: Academic Press.

Verdonk, N. H., J. A. M. van den Biggelaar, and A. S. Tompa, eds. 1983. *The Mollusca, Volume 3, Development*. New York: Academic Press.

Vrba, E. S. 1984a. What is species selection? *Systematic Zoology* 33:318–328.

————. 1984b. Evolutionary pattern and process in the sister-group Alcelaphini-Aepycerotini (Mammalia: Bovidae). In N. Eldredge and S. M. Stanley, eds., *Living Fossils*, pp. 62–79. New York: Springer Verlag.

————. 1989. Levels of selection and sorting. In P. H. Harvey and L. Partridge, eds., *Oxford Surveys of Evolutionary Biology, Volume 6*, pp. 112–168. Oxford: Oxford University Press.

————. 1990. Life history in relation to life's hierarchy. In C. J. De Rousseau, ed., *Monographs in Primatology* 14:37–46.

Vrba, E. S. and N. Eldredge. 1984. Individuals, hierarchies and processes: Towards a more complete evolutionary theory. *Paleobiology* 10:146–171.

Wagner, G. P. 1989. The biological homology concept. *Annual Review of Ecology and Systematics* 20:51–69.

Wiley, E. O. 1978. The evolutionary species concept reconsidered. *Systematic Zoology* 27:17–26.

————. 1981. *Phylogenetics: The Theory and Practice of Systematics*. New York: John Wiley.

Williams, G. C. 1966. *Adaptation and Natural Selection*. Princeton: Princeton University Press.

Wray, G. A. and R. A. Raff. 1991. The evolution of developmental strategy in marine invertebrates. *Trends in Ecology and Evolution* 6:45–50.

Wright, S. 1931. Evolution in Mendelian populations. *Genetics* 16:97–159.

Wynne-Edwards, V. C. 1962. *Animal Dispersion in Relation to Social Behavior*. Edinburgh: Oliver and Boyd.

Index